THE EMPTY RAINCOAT

Charles Handy is an independent writer, teacher and broadcaster, known to many for his 'Thoughts for Today' on the BBC's *Today* programme. He has been in his time, an oil executive, an economist, a professor at the London Business School, the Warden of St George's House in Windsor Castle and the chairman of the Royal Society for the Encouragement of Arts, Manufacture and Commerce. He was named as Business Columnist of the Year in 1994.

Charles Handy was born in Kildare in Ireland, the son of an Archdeacon, and educated in Ireland, England (Oxford University) and the USA (Massachusetts Institute of Technology).

His other books published by Arrow include *Understanding Organizations, The Age of Unreason, Waiting for the Mountain to Move, Beyond Certainty* and *Gods of Management*

He and his wife Elizabeth, a portrait photographer, live in London, Norfolk and Tuscany.

By the same author

Understanding Organizations
Understanding Schools as Organizations
Understanding Voluntary Organizations
The Age of Unreason
Gods of Management
Waiting for the Mountain to Move
Beyond Certainty

THE EMPTY RAINCOAT
Making Sense of the Future

CHARLES HANDY

ARROW
BUSINESS BOOKS

This edition published by Arrow Books Limited in 1995

7 9 10 8 6

First published in Great Britain in 1994 by Hutchinson
Reprinted 1994 (seven times)

Arrow Books Limited
Random House UK Limited
20 Vauxhall Bridge Road, London SW1V 2SA

Random House Australia (Pty) Limited
20 Alfred Street, Milsons Point, Sydney,
New South Wales 2061, Australia

Random House New Zealand Limited
18 Poland Road, Glenfield, Auckland 10, New Zealand

Random House South Africa (Pty) Limited
Endulini, 5a Jubilee Road, Parktown 2193, South Africa

Random House UK Limited Reg. No. 954009

British Library Cataloguing Publication Data

A catalogue record for this book
is available from the British Library

Papers used by Random House UK Limited
are natural, recyclable products made from wood grown in sustainable forests. The manufacturing processes conform to the environmental regulations of the country of origin

ISBN 0 09 930125 3

Printed and bound in Great Britain by
Cox & Wyman Ltd, Reading, Berkshire

Contents

Acknowledgements

There is an invisible corps of people behind this book. They are the individuals who are out there living the lives and driving the organisations which I try to describe. Some of their problems and achievements, their hopes and frustrations, I hear from their own lips in seminars, conferences and private sessions. Some I glean from the writings of others, in journals, newspapers and books. These people must remain anonymous, unless they have chosen to write publicly about their world, but I owe them a debt of gratitude because it is through their stories that I glimpse reality.

I have learnt a lot from the writings of others, be they management theorists, old philosophers or modern thinkers. All those whom I have cited in the text have their relevant works listed in the bibliography at the end. That bibliography also contains some authors not specifically mentioned in the text but whose writings have been particularly influential as I worried about the theme of the book. It is a small recognition of my gratitude to them.

I have had the pleasure of working with two publishers simultaneously, in London and Boston. No one can serve two masters, it was said, but I have found it enormously helpful to be exposed to two sets of views and comments, particularly when they come from such insightful people as Gail Rebuck and Paul Sidey in London and Carol Franco and Natalie Greenberg in Boston. They, and every member of their teams, have been perfect midwives to this book during its rather prolonged birth pangs. I am for ever grateful for their interest in the book, their patience and their encouragement.

My family know all too well the problems of living with a writer. They have been wonderfully tolerant of my moods, have allowed me to parade parts of their lives in the book, and have been tactful critics of the work in progress. My wife's consistent belief in me and in what I am trying to do has been a particular source of strength, seeing me through the valleys of self-doubt, because writing is a lonely business most of the time. To Liz, Kate and Scott go my love and thanks.

Diss, Norfolk, England
September 1993

The Story Behind the Book

Four years ago, my earlier book, *The Age of Unreason*, was published. In that book I presented a view of the way work was being reshaped and the effect which the reshaping might have on all our lives. It was, on the whole, an optimistic view. Since then, the world of work has changed very much along the lines which were described in the book. This should be comforting to an author, but I have not found it so. Too many people and institutions have been unsettled by the changes. Capitalism has not proved as flexible as it was supposed to be. Governments have not been all-wise or far-seeing. Life is a struggle for many and a puzzle for most.

What is happening in our mature societies is much more fundamental, confusing and distressing than I had expected. It is that confusion which I am addressing in this book. Part of the confusion stems from our pursuit of efficiency and economic growth, in the conviction that these are the necessary ingredients of progress. In the pursuit of these goals we can be tempted to forget that it is we, we individual men and women, who should be the measure of all things, not made to measure for something else. It is easy to lose ourselves in efficiency, to treat that efficiency as an end in itself and not a means to other ends.

I cannot forget a sculpture which I saw in the open-air sculpture garden in Minneapolis. It is called 'Without Words' by Judith Shea. There are three shapes. One of them, the dominant one, is a bronze raincoat, standing upright, but empty, with no one inside it. To me, that empty raincoat is the symbol of our most pressing paradox. We were not destined to be empty raincoats, nameless numbers on a payroll, role occupants, the raw material of economics or sociology, statistics in some government report. If that is to be its price, then economic progress is an empty promise. There must be more to life than to be a cog in someone else's great machine, hurtling God knows where. The challenge must be to prove that the paradox can be managed and that we, each one of us, can fill that empty raincoat.

So many things, just now, seem to contain their own contradictions, so many good intentions to have unintended consequences, and so many formulae for success to carry a sting in their tail. Paradox has almost become the cliché of our times. The word crops up again and again as journalists and other writers look for a way to describe the dilemmas facing governments, businesses and, increasingly, individuals. Sometimes it seems that the more we know, the more confused we get; that the more we increase our technical capacity the more powerless we become. With all our sophisticated armaments we can only watch impotently while parts of the world kill each other. We grow more food than we need but cannot feed the starving. We can unravel the mysteries of the galaxies but not of our own families. To call it paradox, however, is only to label it, not to deal with it. We have to find ways to make sense of the paradoxes, to use them to shape a better destiny.

I know precisely when paradox became the key concept in my search for a way to make sense of the confusions. It was in Sausalito, California, when John O'Neil gave me the first chapter of his new book to look at. John is president of the California School of Professional Psychology, a wise and shrewd observer, and counsellor, of leaders and organisations. His new book was called *The Paradox of Success* and was subtitled 'When Winning at Work Means Losing at Life'. The book is about the personal dilemmas of leadership, but the important message for me was that there are never any simple or right answers in any part of life. I used to think that there were, or could be. I now see paradoxes everywhere I look. Every coin, I now realise, has at least two sides, but there are pathways through the paradoxes, if we can understand what is happening and are prepared to be different.

The ideas of *The Age of Unreason* are still relevant, therefore; organisations will become both smaller and bigger at the same time; they will be flatter, more flexible and more dispersed; our working lives will, likewise, have to be flatter and more flexible. Life will be unreasonable, in the sense that it won't go on as it used to; we shall have to make things happen for us rather than wait for them to happen. What I had not anticipated, however, in that first book, was the confusion which this would cause; that the opportunities for personal fulfilment which I so confidently predicted would be complicated by the pressures of efficiency, that the new freedoms would often mean less equality and more misery, and that success might carry a disproportionate price.

One criticism of *The Age of Unreason*, that 'it was all very easy for people like you', did hit home. I am more chary, now, of offering general solutions to our individual predicaments. We must each find our own way. The map, however, will be much the same for all of us, even if we choose to follow different paths. There are pointers to the future in this book, challenges which I think will face all organisations and all individuals, and some frameworks for thinking about them, but, this time, no sure-fire recipes for success.

The important question is whether we shall all be heading in the same general direction. Is there a point to it all, and if so, what is it? Vaclav Havel, the playwright turned president, could hardly be more immersed in worldly things and structures these days, but he has argued that we will only avoid 'mega-suicide' in our time if we rediscover a respect for something otherworldly, something beyond ourselves. It is a paradox, he says, but, without that respect for a superpersonal moral order, we will not be able to create the social structures in which a person can truly be a person. We cannot be the measure of all things, perhaps, unless we have something against which to measure ourselves. I come back to this issue in the last part of the book, but the question of the point of it all is lurking behind every page. The study of philosophy, I was once told, is the study of life, but don't expect it to tell you how to live. A bit like this book, I suspect.

Part One:
In the Dark Wood
Confused by Paradox

1 We Are Not Where We Hoped To Be

It Doesn't Make Sense

There will be no one to pick the olives in parts of Italy this year. The old people are too old and the young will not do it for the money on offer. In Tuscany they did not bother to replace many of the olive groves destroyed in the harsh winter of 1985. It was not worth it. Olive-farming now has to be a serious business, offering serious jobs for serious prices.

Changing, too, are those small family restaurants where the daughter helped her mother in the kitchen and the same waiter was there at lunch-time and in the evening, every day, every week. In most countries the law no longer allows jobs with hours like that, but the result is that eating out is more expensive, like the olives, and many small restaurants are now uneconomic. 'I'm really working for the government now,' said the owner of one, 'collecting their taxes and keeping unemployment down. There is nothing for me at the end of the day.'

We have priced many jobs out of existence all over the industrialised world. People need good salaries or wages to live in these countries. Governments need taxes. Not all products or services can carry these costs. Window-cleaning does not merit a craftsman's wage, nor is a bottle of milk delivered to a British front door each morning really worth more than the price of a bottle of wine. Remove the subsidy and the delivery will end.

Proper jobs are now expensive jobs, providing high-priced goods and services for those who can afford such things. For the rest, it is do-it-yourself, pick your own olives, clean your own windows or collect your own milk. Fair enough. Yet, across a narrow strip of sea from those unpicked Italian olive trees live the Albanians, a people in desperate poverty who would be happy to pick olives or clean windows for a pittance. Every rich country has its neighbouring Albanians. If we let them in to do the work which no one else will do, then someone else will have to pay for their lodgings, their health care and, ultimately, their old age. So we keep them out, mostly, if we can.

Many of them are here already, however. They are our own citizens, but not qualified enough, not perhaps diligent enough, to be able to add more value than the salary or the wage they need to earn in an expensive society. They are, literally, not worth employing in a proper job. Yet they are our citizens, with a right to a life, and, arguably, a right not only to a livelihood but to the sort of work that makes life worth living. They are also the customers for those who are in work. Keep them poor, as potential cheap labour when needed, and you bleed the market of demand. That, at present, seems to be the best that we can do, offering bits and pieces of pocket-money work. America, in the years from 1973 to 1989, managed to create 32 million net new jobs compared with only 5 million in the whole of Western Europe, but it was mostly hamburger work for hamburger pay.

It is one of the dilemmas of a rich society. There are more. Those proper jobs are not unalloyed bliss for all. Much is demanded and expected of those who have them. I asked a young friend, proud of his new job in a London bank, to come for a drink one evening. 'I cannot get away until 9 p.m.,' he said. 'Not ever?' I asked. 'Not really,' he said. 'My group expects me to be there until late, and on most Saturdays too. I can't let them down.' It was exhilarating work, for the most part, he said, and very well paid, but it was totally consuming. His neglected partner said, 'It's a crazy system. It doesn't make sense. Why don't they employ twice as many people at half the salary and work them half as hard? That way they could all lead a normal life.'

But they don't, and they won't and they can't, not if they want to remain competitive. A chairman of a large pharmaceutical company had summed up his policy very neatly once, but it was the other way round – '½ × 2 × 3 = P,' he said: half as many people in the core of his business in five years' time, paid twice as well and producing three times as much, that is what equals Productivity and Profit. Other businesses may not formulate it so crisply but that is the way they are all going: good jobs, expensive jobs, productive jobs, but much fewer of them. It makes good *corporate* sense.

Those jobs are not for everyone. They are not for those who want more space in their lives for other things. For families, for instance. Those kind of jobs are difficult for women if they want to raise a family, or for men, for that matter, who might want to do likewise. Child-rearing can be delegated, of course, but it is not what everyone wants. 'I insist that the company pays for me to read a bedtime story to my children over the international telephone lines when I'm away on business,' said one account-executive mother, but there is more to parenthood than telephoned bedtime stories.

Nor do they last for ever, these jobs. We rightly deplore age discrimination in our societies but 70 hour weeks do wear people out. At some stage, energy must yield place to wisdom, or sometimes just to exhaustion. 'Burn-out' would not have become a popular jargon word if there were nothing for it to describe. We seem, in many of these very full jobs, to be cramming the 100,000 hours of a traditional lifetime's work into 30 years instead of the traditional 47 years, as in days gone by. But then, do we really owe a job to a person who cannot do it any more? Concealed behind those high salaries and big wages is the risk that you may, one day, not be worth it. Sometimes it seems that there is nothing so insecure as a secure job.

A 30-year job leaves 20 years or more 'beyond the job' for nearly everyone, for, if we have not died by the age of 50, we are unlikely to die before 75 unless we do something silly. Those 25 years cannot properly be called 'retirement'. They offer the possibility of another life for all of us. Jung believed that the first half of life is the preparation for the second half. Now that most of us will have the opportunity of that second half in full measure, we are strangely unprepared for it. Many of us waste it. 'All I want is more of the same,' said a friend. Unfortunately, that is seldom on offer.

The dilemmas, and the paradoxes, continue. Akio Morita, the chairman of Sony, has commented that the Japanese worked an average of 2,159 hours each in 1989. That compared with 1,546 hours of the average German. Other countries fell in between. The young Japanese, suggested Morita, will not long tolerate such a divergence, particularly the young well-educated young women who are now joining Japanese corporations. The difference is equivalent, after all, to fifteen 40-hour weeks more than the Germans every year. No wonder, one might think, that the birth-rate in Tokyo is now only 1.1 babies per female, half of what is needed to sustain the population. There is, literally, no time for babies and work. How, and when, these traditions of work will change in such a country of tradition is anybody's guess, but if they do not change, Japan will have an increasingly resentful, ageing and diminishing workforce. Morita's remarks raised the eyebrows of Japan's elders, but, in a 1993 opinion poll, 87 per cent of respondents agreed that they wanted the change.

For Germany, on the other hand, the challenge is to continue to make every hour a German works as effective as one hour and 20 minutes done by a worker in Japan. The

Germans will need to do that in order to maintain their competitive position. It is a demanding standard, even if the Japanese begin to relax. It is a high standard particularly for the now united Germany, where two different traditions of work can still clash.

'Work,' said a friend in Dresden, in the old East Germany, 'used to be a place one went to, not something one did. We could not always work very productively because the parts or the tools we needed were not there. Anyway the customers were used to waiting and we got paid the same whether we did anything or not.' I must have looked appalled because he went on, 'I don't mean that it was right, or even sustainable as a system, but it did mean that there was a lot of time and energy for family and friends, for festivals and fun. Now,' he smiled ruefully, 'it seems to be all about profit and performance, pay and productivity. Sometimes I think that I preferred the four "f"s to the four "p"s! What is it all about?'

We all share, in some degree, the dilemmas of both Japan and Germany. When we worked to ensure our own survival, it was hard but it was understandable. Many are now fortunate enough to be beyond survival. But 'beyond survival' carries with it the question 'What now?' or 'What next?' and a whole variety of answers. They are answers which are increasingly demanded of our political leaders, of our businesses, of our schools and hospitals and prisons and, of course and most pressingly, of ourselves. One way out is to redefine survival. We can define it as keeping up with our neighbours, as individuals, as businesses and as nations. But that has a never-ending, no-win, nightmarish touch to it if we take it seriously. Only one firm can be the industry leader, only one country top economically, there are always richer or more successful neighbours to compare ourselves with. Competition is healthy, maybe even essential, but there has to be more to life than winning or we should nearly all be losers.

Maybe that is already happening. In 1992 the Congressional Budget Office of the United States, a scrupulously non-partisan body, revealed that personal income in the US increased by $740 billion between 1977 and 1989 after adjustment for inflation. Of this total, almost two-thirds went to just 660,000 families, the wealthiest one per cent. For that fortunate group, average income rose from $315,000 to $560,000, or by 77 per cent. The middle classes gained a miserly four per cent over this period while 40 per cent of all families actually ended up worse off in real terms at the end of this decade of affluence. The incentives, which may have been the fertilisers to grow more wealth, ended up consuming all the wealth which they created.

While there are some arguments about the precise interpretation of those figures, it is clear that wealth did not trickle down too well in America during that decade, the Reagan years. Nor did it elsewhere. The figures for Britain are no different. A government report in 1993 revealed that over the period 1979–90, the bottom 10 per cent saw their income in real terms *fall* by 14 per cent, while the average household income *increased* by 36 per cent. The wealth has been slightly less skewed in the other mature economies, but the trend has been the same. As ever, the rich got richer and the poor got, relatively, poorer the world over, and sometimes poorer in absolute terms. What held it all together was only the hope among the poor that, maybe, in a world of constant growth there would be room for some of them, too, amid the rich. It is beginning to seem a rather forlorn hope.

Al Gore, before he became the vice-president of the United States, wrote:

> We have constructed in our civilization a false world of plastic flowers and Astro-Turf, air-conditioning and fluorescent lights, windows that don't open and background music that never stops, days when we don't know whether it has rained,

> nights when the sky never stops glowing, Walkman and Watchman, entertainment cocoons, frozen food for the microwave oven, sleepy hearts jump-started by caffeine, alcohol, drugs and illusions.

He could have made it sound much worse, had he described the wastelands of many inner cities. In these wastelands there are mindless murders of tiny children, rapes of old ladies, burglaries and thefts every 30 seconds in some places, a total disregard for human life and property, senseless anonymous violence.

Al Gore was writing out of concern for the environment. He could as well have been writing out of concern for the human spirit. That we have a spirit, most of us feel sure. We are not incidental curiosities, mutations in the evolutionary process. It would be a waste of all our progress if we sacrificed that human spirit in the pursuit of some imagined efficiency.

Even if we ignore, for a moment, the turbulent conflicts in the old Russian Empire, the endless dilemmas of the Middle East, the pitiless wars and famines of Africa and our continued inability to save what is left of the global environment for our grandchildren, there are enough problems in what we thought were the triumphant capitalist nations to make us wonder if we have missed the road to the future which we thought that we had won.

Some Unintended Consequences of Good Intentions

The millennium is only a statistical accident, but the ending of a thousand years of history does concentrate the mind wonderfully, particularly when it seems to be coinciding with the ending of some things we have taken for granted for the past few generations, such as the employment organisation.

Last Christmas, the family game was to list all the things which had got better in the last decade. The intention was to bring a note of cheer into the proceedings. We all agreed on New Zealand wine and hospices, but got bogged down after that. Some championed CD Walkmans and some the personal phone but these hardly seemed to classify as advancing civilisation. The game soon became too depressing to be fun.

Nevertheless, some things have got better over time. Because of what we have done in the last 50 years, everyone in our industrialised societies now has more things, more equipment, better health and better housing. That must be good news. But these things have their unintended costs and, when we look back, dispassionately, over the last half-century, the news is still mixed. These were the years of my generation, the generation now moving slowly into their Third Age, the age beyond the organisation and full-time responsibilities. It was this generation which set out to build a new world order after the Second World War, which saw capitalism triumph over communism and which kept muzzled the ogre of nuclear war. Some things, however, we did not foresee.

It was this generation which used technology to make a dramatic improvement in productivity, but thought too little about all those who would no longer be required to do the old, essential tasks. What work there will be in future will, for many, be non-essential work, selling goods and services which we could happily do without, building yellow-page economies of glitz and extras, hardly the stuff of real life.

The reward of productivity was increased consumption. To be a customer was seen as the new enlightenment. Even Britain's much-vaunted Citizen's Charter turned out, on inspection, to be a customer's charter. It was not realised soon enough that too much consumption has its costs, that the freedom to drive a car, for instance, ends up too often in the freedom to sit in a traffic jam, or that the delights of tourism dwindle when everyone you meet is also a tourist. We made consumption a measure of achievement, unwittingly creating a society of envy, in which to be poor meant to have less than the average even if the average was quite high.

We misinterpreted Adam Smith's ideas to mean that if we each looked after own own interests, some 'invisible hand' would mysteriously arrange things so that it all worked out for the best for all. We therefore promulgated the rights of the individual and freedom of choice for all. But without the accompanying requirements of self-restraint; without thought for one's neighbour, and one's grandchildren, such freedom becomes licence and then mere selfishness. Adam Smith, who was a professor of moral philosophy not of economics, built his theories on the basis of a moral community. Before he wrote *A Theory of the Wealth of Nations* he had written his definitive work – *A Theory of Moral Sentiments* – arguing that a stable society was based on 'sympathy', a moral duty to have regard for your fellow human beings. The market is a mechanism for sorting the efficient from the inefficient, it is not a substitute for responsibility.

As a result of all the 'progress' of the last 50 years, many have done well, but many not so well, even in the rich societies. In the world at large, the rich still get richer and the poor get poorer in spite of our best intentions. The road we have been on, throughout this century, has been the road of management, planning and control. Those who stood on top of society's mountains could most clearly see the way ahead; they could and should plan the route for the rest and make sure that they follow it. In many ways the bigger the mountain, we thought, the better and clearer the view. We applied this approach to our organisations. We thought this way in government. Even when we said that government should get off the backs of the people we did not really mean it, because the people would then not be managed to their best advantage. We have tried to plan and control world trade and world finance and to make a greener world. There should be a rational response to everything, we thought; it should be possible to make a better world.

It hasn't worked. Management and control are breaking down everywhere. The new world order looks very likely to end in disorder. We can't make things happen the way we want them to at home, at work, or in government, certainly not in the world as a whole. There are, it is now clear, limits to management. We thought that capitalism was the answer, but some of the hungry and homeless are not so sure.

Scientists call this sort of time the edge of chaos, the time of turbulence and creativity out of which a new order may jell. The first living cell emerged, some four million years ago, from a primordial soup of simple molecules and amino acids. Nobody knows why or how. Ever since then the universe has had an inexorable tendency to run down, to degenerate into disorder and decay. Yet it has also managed to produce from that disorder an incredible array of living creatures, plants and bacteria, as well as stars and planets. New life is forever springing from the decay and disorder of the old.

At the Santa Fe Institute, where a group of scientists are studying these phenomena, they call it 'complexity theory'. They believe that their ideas have as much relevance to oil prices, race relations and the stock market as they do to particle physics. In his book about their work, *Complexity*, Mitchell Waldrop describes the edge of chaos as the one place where a complex system can be spontaneous, adaptive, and alive. It is also uncomfortable if you are in the middle of it, as so many of our social institutions are right now.

The Inevitability of Paradox

We need a new way of thinking about our problems and our futures. If the contradictions and surprises of paradox are going to be part of those futures, we should not be dismayed. The acceptance of paradox as a feature of our life is the first step towards living with it and managing it.

I used to think that paradoxes were the visible signs of an imperfect world, a world which would, one day, be better understood by us and better organised. There had to be one proven right way to bring up children, I thought. There should be no reason for some to starve while others gorge. Freedom need not mean licence, violence or even war. Riches for some should not necessarily imply poverty for others. We lacked only the knowledge and the will to resolve the paradoxes. We did not yet know enough about how things worked, but eventually there would be what the scientists call a Theory of Everything and we would, as Stephen Hawking, the Cambridge physicist, put it, perhaps ironically, know the mind of God. In my own sphere, I wrote books which implied that there had to be a right way to run our organisations and our lives, even if we could not yet be completely sure of what it was. I was in the grip of the myth of science, the idea that everything, in theory, could be understood, predicted and, therefore, managed.

I no longer believe in a Theory of Everything, or in the possibility of perfection. Paradox I now see to be inevitable, endemic and perpetual. The more turbulent the times, the more complex the world, the more the paradoxes. The Theory of Complexity has been added to the Theory of Chaos. The turbulence, the theory goes, is a necessary prelude to creativity and some new order. We can, therefore, and should, reduce the starkness of some of the contradictions, minimise the inconsistencies, understand the puzzles in the paradoxes, but we cannot make them disappear, nor solve them completely, nor escape from them, until that new order becomes established. Paradoxes are like the weather, something to be lived with, not solved, the worst aspects mitigated, the best enjoyed and used as clues to the way forward. Paradox has to be *accepted*, coped with and made sense of, in life, in work, in community and among the nations.

There was, I now recall, a small framed printed motto which hung in my boyhood bedroom: 'Life goes, you see, to golf's own ditty: Without the rough there'ld be no pretty.' I have no idea why it was there. My family did not go in for such things and my mother had probably picked it up at some charity bazaar. Accidental or not, it was my first subliminal introduction to the necessity of paradox in human affairs. As I grew older, I realised that what I was told had been God's great gift to mankind – choice – was itself a paradox, because the freedom to choose implies the freedom to choose wrongly, to sin. You cannot have the one without the other. Original sin is the price we pay for our humanity. There was paradox at the centre of religion. Quite right, too, I came to realise, because paradox is what makes life interesting. If everything was an unmixed blessing, life would soon begin to cloy. There would be no need for change or movement. Offer me a heaven without paradox and I will opt for hell. Perfection, then, is neither possible nor, perhaps, desirable.

That conclusion was, for me, a revelation. Life will never be easy, nor perfectible, nor completely predictable. It will be best understood backwards, but we have to live it forwards. To make it liveable, at all levels, we have to learn to use the paradoxes, to balance the contradictions and the inconsistencies and to use them as an invitation to find a better way. Scott Fitzgerald once said that the test of a first-class mind was the ability to hold two opposing ideas in the head at the same time and still retain the ability to function. If he was right, then we are in for a time of paralysis, because there are not that many first-class minds around.

Schumacher also put it well: '[Some people] always tend to clamour for a final solution, as if in life there could ever be a final solution other than death. For constructive work, the principal task is always the restoration of some kind of balance.'

Living with paradox is not comfortable nor easy. It can be like walking in a dark wood on a moonless night. It is an eerie and, at times, a frightening experience. All sense of direction is lost; trees and bushes crowd in; wherever you step you bump into another obstacle; every noise and rustle is magnified; there is a whiff of danger around; it seems safer to stand still than to try to move. Come the dawn, however, the path is clear before you; the noises are now the songs of birds and the rustle in the undergrowth is only scuttling rabbits; the trees define the path instead of blocking it. It is a different place.

Prophets and Kings

'There are kings and prophets, I was always told,' said Tony Benn, the British socialist politician. 'The kings have the power and the prophets have the principles.' I am on the side of the kings, the people who make things happen, but every king needs his prophet, to help him, and increasingly her, keep a clear head amidst the confusions. No one, however, would want the prophet to run the show.

Prophets, in spite of their name, do not foretell the future. No one can do that, and no one should claim to do that. What prophets can do is to tell the truth as they see it. They can point to the emperor's lack of clothes, that things are not what people like to think they are. They can warn of dangers ahead if the course is not changed. They can, and often did, point their fingers at what they thought to be wrong, unjust or prejudiced. Most of all, they can offer a way of thinking about things, a way to clarify the dilemmas and concentrate the mind.

What the prophet cannot, and should not, do is to tell the doers what to do. That would be to take the power without the responsibility, the prerogative of the harlot, they used to say, not the prophet. It would be to steal other people's decisions. The prophet can provide a chart but cannot dictate where or how the vessel should sail.

It is my hope that this book will make it easier for people to see their way through the confusions of our times. Some of those people will be the leaders and executives of our institutions, because, in what I see as the ending of the age of the organisation, those institutions will have to find for themselves very different futures. Yet, in their new forms, they will be more essential than ever.

Some of the people reading this book will be individuals, trying to make some sense of their lives. Young people, in particular, face a world very different from the one their parents grew up in, a world where there are not many models from the past for them to draw on, where they really do have to reinvent their lives, their purposes, their standards and their priorities.

Lastly, I would like to think that the ideas and thoughts in the book might be useful to those responsible for the governance of our society, at all levels. They have the awesome responsibility of finding a structure for a society where most of the ground rules have changed but where the need for justice between groups, and between the present and the future, is greater than ever.

There is a need for a new perspective on life, on its purpose and its responsibilities. There are few great causes or crusades any more. Maybe it is the end of history. Some people are cocooned in comfort, others in poverty; but, for either group, their own survival seems to be the only point of life. If that is so, we shall all lose in the end. If anything is to happen, however, it has to start with us, individually, in our own place and time. To wait for a leader to guide us into the future is to be forever disillusioned.

2 The Paradoxes of Our Times

If we are to cope with the turbulence of life today, we must start by finding a way to organise it in our minds. Until we do that we will feel impotent, victims of events beyond our control or even our capacity to understand.

Framing the confusion is the first step to doing something about it. Analysts and therapists know this, of course, but so do the teachers of managers and of doctors. The management students at my business school are regularly confronted with 30-page case-studies, descriptions of the state of play in a business or an industry. This is not a facile attempt to give them an illusion of reality, but a way of teaching them that the first thing to do when confronted with a load of data, impressions and confused signals is to put them into some sort of framework, rather as the doctor learns to turn symptoms into a diagnosis. Only then can treatment begin.

I have identified nine principal paradoxes, nine ways of explaining what is going on in our societies and why some confusion is inevitable. 'Shall life succeed in that it seems to fail,' wrote Browning, '[is] a paradox which comforts while it mocks.' Simultaneous opposites are the other feature of paradox, as when we find that we can, at times, dislike those whom we love the most, but go on both loving and disliking. Paradox does not have to be resolved – only managed.

It is far from being an exhaustive list, but if we can manage the following nine paradoxes, if we can make sense of them, if we can combine their unexpected twists and their contradictions to forge a better world, we shall have done well. They are the paradoxes of the mature economies. Not all of them are yet to be found, for instance, in South-East Asia, still less in Africa. But their time will come, for these paradoxes seem to be the companions of economic progress everywhere.

1 The Paradox of Intelligence

In January 1992, Microsoft's market value, for a time, passed that of General Motors. The *New York Times* commented that Microsoft's only factory asset was the imagination of its workers. Tom Peters proclaimed the symbolic end of the Industrial Revolution. Peter Drucker heralded the post-capitalist society. This may, of course, be a bit premature – imagination is fragile, and Microsoft, it is already clear, should not be complacent. But organisations and individuals everywhere are waking up to the fact that their ultimate security lies more in their brains than in their land or their buildings. Even in the beleaguered world of American auto-making, brains are replacing brawn. In Ford's new Atlanta plant, each car needs only 17 hours of direct labour. Clever workers with clever machines have put an end to the mass organisation.

For a long time now, corporate chairmen have been saying that their real assets were their people, but few really meant it and none went so far as to put those assets on their balance sheet. That may change. Peter Drucker points out that the 'means of production', the traditional basis of capitalism, are now literally owned by the workers because those means are in their heads and at their fingertips. What Marx once dreamt of has become a reality, but in a way which he could never have imagined.

Focused intelligence, the ability to acquire and apply knowledge and know-how, is the new source of wealth.

Singapore, which calls itself the Intelligent Island, recognises in its latest plan that the traditional sources of wealth and comparative advantage – land, raw materials, money, technology – can all be bought in when and if needed, *provided* one has the people with the intelligence and the know-how to apply them. Singapore, along with Hong Kong, has exported all its manufacturing activities to cheaper places like Sumatra, the Philippines or Guandong in China, but retains the managerial control, the design and the distribution – the intelligence quotient.

What is true for Singapore is true everywhere. The new source of wealth in our societies is the intelligence quotient. Intelligence is the new form of property. Unfortunately, it does not behave like any other form of property, and therein lies the paradox. It is, for instance, impossible to give people intelligence by decree, to redistribute it. It is not even possible to leave it your children when you die. You can only hope that there is some of it in their genes. Of course, there is education – which becomes the crucial key to future wealth – but it is a key which takes a long time to shape and a long time to turn. The situation gets odder. Even if I do manage to share my intelligence or know-how with you, I still keep it all. It is not possible to take this new form of property away from anyone. Intelligence is sticky.

Nor is it possible to own someone else's intelligence. Peter Drucker is right – the means of production can, in practice, no longer be owned by the people who think they own the business. It is hard to prevent the brains walking out of the door if they want to. Buying shares in Microsoft is a bet that the imagination of its workers continues to be exercised on Microsoft's behalf and that the imagination never flags. It makes an insecure basis for the stock-market. Intelligence is a leaky form of property.

An added complication is that intelligence is extraordinarily difficult to measure, which is why intellectual property seldom appears on balance sheets. But this also makes it difficult to tax, unlike any other form of property, which makes any form of wealth or property taxation ineffective. Intelligence is tricky, as well as sticky and leaky.

The better news in all of this is that while it is impossible to redistribute intelligence by administrative fiat, it is also impossible to stop people getting it. Anyone can, in theory, be intelligent in some way or can get intelligent, and thereby have the access to power and wealth. There is little to stop any small firm muscling in on Microsoft's territory just as Microsoft did to IBM. Where the key property is intelligence you do not have to be big or rich to get in on the act. It is a low-cost entry market-place. It should make for a more open society.

Unfortunately, intelligence tends to go where intelligence is. Well-educated people give their families good education, which gives them access to power and wealth which, in turn, gives them advantage in the educational market-place for their children. The most likely outcome of the new form of property, therefore, is an increasingly divided society unless we can transform the whole of society into a permanent learning culture where everyone pursues a higher intelligence quotient as avidly as they now look for homes of their own. A property-owning democracy is an exciting thought with this new definition of property.

As a small indicator of the changing perception of property, we may observe that the richer we are the less need we seem to feel to own our own homes. In Bangladesh over 90 per cent of houses are owner-occupied, in Ireland 82 per cent. Go to rich Germany, the Western part, and the figure falls to 45 per cent. In richer Switzerland it is only 33 per cent. Where brains prevail, security lies not in the physical property but in the intelligence quotient. There are, then, better uses for one's cash than buying houses.

2 The Paradox of Work

We all need something to do. Activity is natural. It is hard to see why there should be any shortage of it, yet enforced idleness seems to be the price we are paying for improved efficiency. Why should we worry? To be pleasurably idle was the dream of the ancients, their concept of civilisation. 'If work were so great,' quipped Mark Twain, 'the rich would have hogged it long ago.' They have, Mr Twain, they have. The result is that some have work and money but too little time, while others have all the time but no work and no money. Those with the privilege of idleness see it as a curse because they tend to be at the bottom, not the top, of the heap. We seem to have made work into a god and then made it difficult for many to worship.

Why has work become so lumpy? Part of the problem is money. Work is society's chosen way of distributing income. We will do even boring work for the money it brings. Therefore it would be convenient if everyone had some work to do, even if it was boring work, as a way of getting the money to them. This was part of the communist philosophy. Unfortunately, we also use money as the measure of efficiency. Our organisations, therefore, want the most work for the least money while individuals typically want the most money for the least work. In a competitive world where everything is traded, it is not hard to see that the organisation is going to win.

Organisations are responding to the challenge of efficiency by exporting unproductive work, and people, as fast as they can. Instead of keeping a pool of slightly surplus labour and skills inside the organisation as a sort of cushion for emergencies and comfort, they are pushing those skills outside and pulling them in when necessary. If you approve of this you call it 'getting rid of slack', if you disapprove of it you talk of 'exporting their flexibility on to the peripheral labour market'. Put many of the full-time workers outside and it is they, not the organisation, who will stand the costs of their unused time. Slack always costs money. It is only a question of who pays for it.

The irony is that these unused workers still have to have some money if they are to live and to enjoy some of the rights and pleasures of citizenship. The money has, ultimately, to come in some way from the organisations they left, usually in the form of higher taxes. In the end, much the same work gets done, for total output in the economy has not risen that much, and much the same money gets paid out, but in different ways. In theory it need not happen like that. In theory, those spare workers with their spare time will invent new work to keep themselves busy and in cash. Unfortunately, they are usually the people least capable of creating new work for themselves, because they lack the kinds of intelligence and inclinations which would allow them to be independent. Conditioned to life as employees, we now expect them to be entrepreneurs.

To herald the New Year of 1993, Burton, the British chain of clothing shops, announced that they would be cutting 2,000 full-time jobs but creating 3,000 part-time ones. It was, they said, a strategic response to the stretched day and the stretched week of modern retailing. They were typical of a trend. It is no wonder, then, that only 55 per cent of Britain's workforce is now in full-time employment. There is a lot of spare capacity in the economy these days, but it is in individuals not organisations. It is not clear how we unlock it, except by giving these outsiders a share in the new property – the intelligence quotient. Until we do, re-lumping the work, so that some have too much and some too little, will only divide society.

In fact, Britain and the US have the most open labour markets and, therefore, the highest number of people in work – but their workers are the least protected and often the worst paid. Some 70 per cent of working-age Americans and Britons are in paid work. That compares with 60 per cent in France and only 50 per cent in Spain. Over the last 20 years the numbers of people in paid work have grown by 30 million in America but only by 10 million in the European Community. But the Americans and British have to work longer and odder hours, accept more part-time work and self-employment and enjoy less protection. Fifteen per cent of British workers put in more than 48 hours a week and 20 per cent regularly work on Sundays. The Continentals think this mad. Britain and America add on less than 30 per cent to the wage or salary to take care of social security and pensions. Italy, France and Germany add 50 per cent. Should you have fewer, better-paid, better-educated and better-protected workers, or more but cheaper ones? The argument rages, with the continental Europeans arguing that only good, and therefore expensive, labour is tolerable and worthwhile in this modern age, and that no work is better than bad work. The British and Americans believe that any work is better than no work, even if the result is a progressively downskilled workforce. One consequence of that is a more divided society. In America the top 10 per cent of earners are paid six times as much as the bottom 10 per cent. In Germany the ratio is just over two.

Work is more than a job. There are more forms of activity than paid work. Indeed, if work is priced at zero there is unlimited scope for it. 'I know that only too well,' my housewife friends respond. 'If we were paid for what we do, much of what we do would not be worth doing. We could not afford the price of a clean house or an evening meal if that price had to include our reasonable wages.' If, therefore, people want work for reasons other than money – for self-respect or identity, to make a contribution or to feel part of something – the answer is to price more work at zero. In a society where a lot of the work is not priced everyone is busy. Go to China or any developing country if you want to see bustling activity. Ironically, the more you price work, the less paid work gets done, because so much of it is not now worth the cost. Any work which is worth its price is quickly turned into a business, where few are paid well rather than many paid badly, because that way efficiency lies. Perhaps we should only have work which is priced expensively and work which is priced at zero, rather than fiddling around in between. That, however, leads to the paradox of productivity.

3 The Paradox of Productivity

Productivity means ever more and ever better work from ever fewer people. That is good for the customer and good for the organisation, be it a business or any public service. No one has ever been against efficiency. It has generally, in the end, been good also for the workers, even those who are not included in the 'fewer'. The ones who stayed got better jobs, and better-paid ones. Those who left found work in other growing organisations. Over time they moved into the new growth sectors in the economy. Thus it was that the agricultural workers 200 years ago began to find new work in the new factories, and their descendants (when the factories started slimming and closing) moved to the offices and shops of the service sector. Growth and work went on. As long as the overall growth rate was at least equal to the rate of improvement in efficiency, plus the rate of growth of the population, there would always be jobs somewhere for everyone.

This time, however, the new growth sector for work is the do-it-yourself economy. Some of that do-it-yourself economy is paid for and counted, being the self-employment sector which is growing everywhere; some of it is paid for but not counted – the black economy; some is the purely destructive do-it-yourself of drugs and theft and violence. Much of this do-it-yourself economy is, however, neither paid for nor counted nor illegal, as when we look after our old and sick, do our own repairs, grow our own food. As more and more people get pushed out or leave organisations, it makes good economic sense for them to do for themselves what they used to pay others to do for them. They should, logically, go in for a little personal import substitution, something which every government advocates as desirable on a national basis, but would rather that we did not do individually and domestically. Why pay other people to do or make what you can do or make yourself if, now, you have more time than money on your hands? Because this new growth sector is invisible, productivity does not seem to be producing the output increases, nor the conventional jobs, which we would have expected.

This is not a temporary paradox, governments and the unemployed please note. Society, and individuals, will have to get more used to the do-it-yourself economy as the new growth sector. More of us are going to be in it, whether we like it or not. Better technology means that more and more of us can run businesses or services by ourselves. More of us will be outside the organisation and the formal economy. The OECD calculated that, in 1992, only 33 per cent of the British aged over 55 were in paid work. Before you write that off as a British peculiarity you should know that the figure for France was 27 per cent and for Italy 11 per cent. The rest were not all doing nothing, but what they were doing was not counted, statistically or, more crucially, socially.

Economies have traditionally grown, in measured economic terms, by turning unpriced work into priced work, because then that work can be counted. The irony is that you may then actually reduce the work that is done, although the economy appears to grow. By pricing the work, we turn 'activity' into 'jobs' and so create employment, but some work then gets too expensive for the customers to afford and so on longer gets done. In many cases, it can't be done by ourselves for free because we have forgotten how to do it. The activity disappears. By pricing work we can destroy work, but we will never notice it because it never got counted in the first place.

My friend used to grow all his own vegetables; it was a source of pride to him that he ate for free, even producing his own seeds. The visible economy was the poorer for it, because nothing was bought or sold. As he grew older and richer, however, he calculated that this was a poor use of his time. He would be better off spending more time on his own work and buying the vegetables in the supermarket. The visible economy grew a notch or two thereby. But my friend lost his job and could no longer afford to buy any but the cheapest vegetables. Unfortunately, by this time he had disposed of his vegetable plot and all his tools. He had no energy to start all that again. He was bored, poor and hungry. The economy had slipped back a notch again, but now less vegetables in total were being consumed in that house, there was more idleness and more dissatisfaction. By pricing his work my friend had ultimately destroyed his work.

His story was a parable of the rich societies, who by pricing work have increasingly drawn more types of work into the formal economy. By so doing they have encouraged specialisation and efficiency but have then, as a result, priced some of that new work out of existence, deskilled many of its citizens and created a class of people who have nothing to do if they have no job. It is all an unintended consequence of good intentions, a fall-out from progress, but it is one of the more uncomfortable paradoxes of modern times.

4 The Paradox of Time

In this turbulent world we never seem to have enough time, yet there has never been so much time available to us.

We live longer, we use less time to make and do things as we get more efficient, and should therefore have more time to spare. Yet we have made this strange commodity into a competitive weapon, paying over the odds for speed. If we were wise would we not take the price-tag off time, and give ourselves time to stand and stare?

There was a time when we knew what time was. Patricia Hewitt of Britain's Institute of Public Policy Research has put it neatly – the time that men spent on paid employment determined how much time they had for their families; the time women spent caring for their families determined how much time they had for work. Most men spent most of their time in or around organisations. Most women spent most of their time working in the home. Organisations, you might say, were organised for male convenience, but, as a result, time was more or less fixed. We all knew who was where and when.

I have been using the past tense. Only one-third of British workers now work the 'normal' nine-to-five day, give or take an hour or two at each end. The normal is now the minority. Time is coming unfixed. Organisations want more flexibility. We have to rethink time and the words that we have come to attach to time. I can see a situation coming in society when it will no longer be possible to draw a hard distinction between full- and part-time work, when 'retirement' will become a purely technical term, indicating an entitlement to financial benefits, and when 'overtime' as a concept will seem as outmoded as 'servant' does today. At present, however, time is more unbalanced than balanced for many, which means that their life is also out of balance. Some have more time than they know what to do with, while others have too little time to do all they want to.

Organisations are now rethinking time for their own advantage. It is as if they had finally realised that there are actually 168 hours in the week, not 40. Sleeping assets make no money, so why keep them shut down for 128 hours a week when half of the world is still awake, when customers like to shop at the end of days and the end of weeks, and when some people like to work when others sleep? Most factories are now like processing plants, working the 24-hour day. There are night shifts in financial offices, stores in London stay open until 9 p.m. or 10 p.m. and on Sundays. Schools in Wandsworth in South London have abandoned the long summer holiday, originally designed to allow pupils to help with the harvest, in favour of five 8-week terms. No time demarcations are sacrosanct any more.

There is now a long list of the ways in which organisations are re-chunking time. There is flexitime, which has been with us for a while now, but if we moved to a 35-hour week, flexitime could mean an hour off each working day, or Friday afternoons off, or a nine-day fortnight. Then there is part-time working for new parents, part-time before retirement, job-sharing, term-time jobs, weekend jobs, four 10-hour days a week or eight-day fortnights, annual-hours contracts, zero-hour contracts (being available as and when required), parental leave, career breaks, sabbaticals, time-banking (accumulating holiday entitlements over several years), and individual-hour contracts where individuals and their bosses agree on a timetable of hours each week or each month.

On the face of it, there is enough flexibility available for everyone. Why, then, does Juliet Schor need to write a book called *The Overworked American* which sells so well that it must have struck a chord with many? The average American, she finds, now works 164 more hours per year than 20 years ago – the equivalent of an extra month. The typical American now works 47 hours per week and, if current trends continue, in 20 years the average person would be on the job 60 hours a week, for an annual total of 3,000 hours. That compares with 1,856 in Britain in 1989. Why do they do it? Schor says that two things come together:

organisations want fewer people working longer because it saves them overheads; while individuals want the money. This 'Faustian bargain of time for money', says Schor, has created an insidious cycle of work and spend, as people increasingly look to consumption to give satisfaction and even meaning to their lives.

The paradox is that they seem to know that it is stupid. In a US Department of Labor survey in 1978, 84 per cent said that they would choose to trade off some future increases in income against more time, with almost half opting to trade all the increases. In Britain, Andre Gorz recorded the overtime-loving workers at a shoe factory. When hard times hit, the factory went into work-sharing, and the employees who had chased all the extra hours they could get – including Sundays and holidays – now found themselves with time on their hands. One worker reported:

> Bit by bit, there was an unbelievable phenomenon of physical recuperation. The idea of money really lost its intensity. It's quite true that we lost a good deal of money [25 per cent of previous income] but, quite soon, only one or two of the blokes minded. It was about now that . . . friendships began: we were now able to go beyond political conversation, and we managed to talk about love, impotence, jealousy, family life . . . it was also at this time that we realized the full horror of working in the factory on Saturday afternoon or evenings . . . we were once again learning the meaning of living.

Schor says that we seem to have decided to take the benefits of the improved productivity of the last 50 years in money rather than time. Work and spend has become a habit. She is the first to recognise, of course, that for some people there is no choice. Nearly one-third of American workers earn wages which, on a full-time basis, would not lift them out of poverty. The same is true of Britain.

Millions of households can only make ends meet through overtime, moonlighting or multi-earning households. They would willingly give more time to make more money, just to live, just to make ends meet.

The trouble started when we turned time into a commodity, when we bought people's time in our organisations rather than buying their produce. The more time you sell, under these conditions, the more money you make. There is, then, an inevitable trade-off between time and money. Organisations, for their part, get choosy. They want less time from the people they pay by the hour but more from the people they pay by the year, because, in the latter case, every extra hour during the year is for free.

Time turns out to be a confusing commodity. Some will spend money to save their time, others will spend their time to save money. Others, again, will trade money for time at certain periods of their life, preferring to work less long for less money. This makes time a contradictory sort of commodity but one which will become more and more important in our societies.

Busy people will, if they can afford it, spend money to save time, buying time-saving equipment for their homes, pre-cooked meals and help with their chores; they will prefer taxis to buses, child-minders to child-minding, gardeners to gardening, if it helps them to spend their time on what they really want to do. Their needs create an important market opportunity. The unbusy, on the other hand, spend money to buy time – time to travel, time to learn, time to play and time to keep fit – or they spend their time doing themselves what they used to pay others to do, in the do-it-yourself economy. Time, therefore, creates the new growth area. Personal services for the busy, to save time; health, education, travel and recreation for the affluent unbusy, to spend time; equipment and materials for those who want to spend time to save money. It is, perhaps, no accident that these new growth areas will not be best served by large corporations but by small independents providing a personal and local delivery, linked maybe by franchising or other networks into bigger combinations.

5 The Paradox of Riches

Economic growth depends, ultimately, on more and more people wanting more and more of more and more things. Looking at the world as a whole, then, there should be no shortage of growth potential. If, however, we look only at the rich societies, we see them producing fewer babies every year and living longer. Fewer babies mean fewer customers, eventually, while longer lives mean, usually, poorer and more choosy customers. Older people, even when they have the money, are in a slimming-down, passing-on stage, not a stocking-up one. We could be running out of customers at home.

But not abroad; not, anyway, in the multiplying needy areas of the developing world. They, however, cannot afford to buy most of the things we have to sell. What they want is the know-how and the capital to make things to sell to us, before they can start to buy from us. Ultimately, therefore, we will have to invest in our potential competitors in order to fuel our own growth. No government has been able to persuade its people to accept that paradox, although the multinational businesses are beginning to see the sense, for their shareholders, of making things where they are cheaper, wherever that is, and exporting their know-how to make that possible. In the short term, however, exporting factories, and know-how not products, is nothing but bad news for those who used to work in the factories back home. It is their children who will benefit from a richer world outside, not them. Will they be prepared to make the sacrifice?

Back home, the traditional answer has been to create

even more demand among those who do have the money. Growth has to be fuelled by what the American economist, Thorstein Veblen, first called 'conspicuous consumption' 100 years ago, the need to be up with the neighbours or better than them. Growth, then, which is necessary for society, is increasingly dependent on a climate of envy in that society, increasing its divisions. Paradox again. There are, however, some signs that the 'Gucci factor', the high-fashion luxury trade which is based on envy, may have peaked in the Eighties, along with the firm of that name. It is, said the *Financial Times*, 'the demise of de luxe'. Couture houses in Paris are worried that no one will want to pay the prices for their creations. Consumers have become more discerning, less interested in conspicuous consumption, asking more often 'Will it work well?' or Will it last?' We have, paradoxically, to wonder whether this is good news or bad. It is bad for growth, good for common sense.

We shall miss one customer, the artificial customer of the defence industries of the West. Politically legitimate, the defence industry created a demand for advanced technology which spread knowledge and work throughout the economies of America, Britain and most Western countries. In yet another paradox, one must hope that this customer will never again be needed to the same extent. For the good of our economies, however, an alternative, and politically legitimate, artificial customer would be very beneficial. We could, for instance, define the environment as the new target for defence expenditure, fending off our own deterioration. Sadly, turning swords into ploughshares has never proved easy. Peace dividends quickly disappear into the national loan account.

6 The Paradox of Organisations

We used to think that we knew how to run organisations. Now we know better. More than ever they need to be global and local at the same time, to be small in some ways but big in others, to be centralised some of the time and decentralised most of it. They expect their workers to be both more autonomous and more of a team, their managers to be more delegating and more controlling. The paradox is neatly summed up in Charles Savage's story, in his book *Fifth Generation Management*, of the manager saying to the new recruit, 'The good news is that you have 120,000 people working for you, the bad news is that they don't know it.'

John Stopford and Charles Baden-Fuller, in their study of rejuvenating businesses, report that the successful ones live with paradox, or what they call dilemmas. They have to be planned and yet be flexible, be differentiated and integrated at the same time, be mass-marketers while catering for many niches; they must introduce new technology but allow their workers to be the masters of their own destiny; they must find ways to get variety and quality and fashion, and all at low-cost; they have, in short, to find a way to reconcile what used to be opposites, instead of choosing between them.

Charles Hampden-Turner, in his book on corporate culture, also focuses on the inevitable dilemmas of organisations, arguing that managers have to be 'masters of paradox', turning the horns of the various inevitable dilemmas into virtuous not vicious circles. He quotes, as an example, the Berkeley consultants, Meridian, who use the mythical Greek image of Scylla, the rock, and Charybdis, the whirlpool, which Odysseus and his sailors had to steer between, to characterise the hard and soft features of organisations, the structured, controlled, masculine side and the flexible, responsive, feminine side, both of which are needed for success.

These authors speak as if we would know an organisation when we see it, full of paradox though it may be. The organisations of the future may not be readily recognisable as such. When intelligence is the primary asset the organisation becomes more like a collection of project groups, some fairly permanent, some temporary, some in alliance with other parties. Instead of an organisation being a castle, a home for life for its defenders, it will be more like a condominium, an association of temporary residents gathered together for their mutual convenience. The condominium may, in fact, not have any physical existence, because the project groups or clusters do not have to be all in the same place. This has caused some to talk of the 'virtual corporation', something that can be discerned more easily on the computer screen than in the physical world. The challenge for tomorrow's leaders is to manage an organisation that is not there in any sense which we are used to.

It is, however, a challenge that must be won, because these minimalist, partly unseen, organisations are the linchpins of our world. We may, most of us, not belong to them, but we shall be selling our services into them; the wealth of our societies will depend upon them; they will, ultimately, be the source of our well-being. The age of the organisation may be coming to an end in one sense, when to be a full-time employee is a minority occupation, and when even for this minority the time in the organisation represents less than half one's adult life, but in another sense the organisation, or what is left of it, will be the critical component of society. Organisations will organise, but to do so they will no longer need to employ. An organising organisation will look and feel very different from an employment organisation. Because it will be less visible as an organisation we should not think it is less important.

7 The Paradox of Age

We all age, but each generation ages differently. This is technically called the cohort factor. Each cohort or generation is affected by its own history. It is, therefore, very unlikely than my children will have the same sort of life-cycle as I have had, or that mine would be like my parents'. My parents' generation saw a world war, in some cases two world wars. They went through a long and deep recession in the Thirties. They valued security above all else and they expected to, and did, work until almost the end of their lives.

Society expected that it would be the same for their children. It wasn't. Work inside organisations has petered out for many in their fifties, creating the kind of mid-life crisis for many which their parents had never known. Change speeded up. The world got smaller. Children did not die and were not killed in wars with the result that we planned for smaller families, one child, sometimes, in place of three. Divorce replaced death as the end of many a marriage, creating the kind of spreadeagled families which to our parents were a rarity. The problems which we encountered were new. The crises were different. Society, however, was still geared to the ageing patterns of the generation before. Pension schemes, divorce laws, social expectations were all inappropriate and took time to change.

It will happen again. My children, unlike previous cohorts, will find conventional jobs and careers harder to get. Their work lives will start later and end earlier, creating a gap between adolescence and adulthood which their parents never knew, a gap which, therefore, they, and we, do not really know how to fill. Their relationships will be different again from ours. Because they have grown up without wars they will be more carefree with their plans and their lives. Their education will have to be more prolonged, if not indefinite. Women will all do paid work for most of their lives but both sexes may want and need to find intervals of child-rearing and learning. Children are now a decision not an accident. The roles of the sexes will change, and bring with that change different values and priorities.

The paradox of ageing is that every generation perceives itself as justifiably different from its predecessor, but plans as if its successor generation will be the same as them. This time it needs to be different.

8 The Paradox of the Individual

Society speaks with two voices. One voice urges us to discover our 'authentic self', to be ourselves, to plan our own path through life and, whilst respecting the rights of others, to hold fast to the right to be true to ourselves. Individualism acquired a bad name in the years of Reagan and Thatcher when it was used to justify the uninhibited pursuit of private gain, masquerading as 'enterprise'. There was once, however, a more honourable British tradition, drawing on Darwin's idea of self-reliance and a rich tradition of eccentrics and opinion-leaders. It was then, and is now again, respectable and desirable to be yourself.

The other voice is that of the receptionist or the conference-organiser. 'Who do you represent?' 'To whom are you affiliated?' 'What organisation are you from?' Recounting his problem with receptionists and switchboards, the British writer, Anthony Sampson, who works on his own from home, says, 'I'm tempted to reply that I represent the human race . . . the inalienable right to life, liberty and the pursuit of happiness, but it won't get me through the switchboard. I have to reply that I represent no one or that "I'm just a friend". I feel even more freakish at conferences where everyone else seems to represent some company, organisation or group.' It was, he points out, John D. Rockefeller, the creator of Standard Oil, the first modern corporation, who remarked, 'The day of combination is here to stay. Individualism has gone, never to return.' He was only partly right. An MIT study comparing American and Japanese working methods, did conclude that the individualism of American workers had to be balanced by the combination of a good team if they were to match the productivity of the Japanese. The Japanese, however, are looking for some of that individualism and creativity to balance the conforming force of their combinations.

It is a paradox, one best captured by Jung, who said, years ago, that we need others to be truly ourselves. 'I' needs 'We' to be fully 'I'. Looking up, however, at the office-blocks in every city, those little boxes piled on top of each other up into the sky, one has to wonder how much room there is for 'I' amid the filing cabinets and the terminals. It was A. E. Housman, Sampson reminds us, who wrote in one of his poems, 'I, a stranger and afraid in a world I never made.' Who, we must wonder, will be the 'We' to whom we would want to belong? Is it the minimalist, virtual organisation? Or our current 'edge city' in suburbia? Or the disappearing family? Can a personal network substitute for these?

9 The Paradox of Justice

Justice is the bond of society. We are happy to belong to a society which treats us fairly, which gives us our due and which is impartial. The problem is that 'giving each their due' can mean a variety of contradictory things. It can, for instance, mean giving us what we deserve, be that a reward for achievement or a punishment for offences. On the other hand it can mean giving us what we need. Political parties will champion one or the other definition and claim to be the party of justice. Both will be right.

Michael Young, writing 30 years ago, summed up the dilemmas of distributive justice very nicely:

> One could say that it was wrong to pay one man more than another because there should be distribution according to needs.

> One could say that it was wrong to pay the lazy scientist more than the diligent dustman because there should be distribution according to effort. One could say that it was wrong to pay the intelligent more than the stupid because society should compensate for genetic injustice. One could say that it was wrong to pay the stupid more than the intelligent because society should compensate for the unhappiness which is the usual lot of the intelligent. (No one can do much about the brilliant, they will be miserable anyway.) One could say that it was wrong to pay the man who lived a long and serene life in Upper Slaughter as much as a scientist who wore himself out in the service of knowledge. One could say that it was wrong to pay people who liked their work as much as those who didn't. One could – and did – say anything, and whatever one said it was always with the support of . . . justice.

Thirty years on, the dilemmas remain. Justice, argue some, needs to treat everyone fairly, that is equally – unless there are very good arguments for unequal treatment. That is fair to the underdog but is less fair to those who, perhaps, 'deserve' more, because they contribute more. What is clear is that a society which is perceived to be unjust will earn no loyalty or commitment from its citizens; there will be no good reason for anything other than selfishness. Such a society is doomed, in the end, to destroy itself.

Capitalism thrives on the first definition of distributive justice – those who achieve most should get most. But it will not long be credible or tolerated if it ignores its opposite, that those who need most should have their needs met. To put it another way, capitalism depends on the fundamental principle of inequality – some may do better than others – but will only be acceptable in the long term in a democracy if most people have an equal chance to aspire to that inequality. It is a paradox which we cannot afford to ignore.

Part Two: Finding the Balance
Pathways Through Paradox

Paradox confuses because things don't behave in the way we instinctively expect them to behave. What worked so well last time around is not guaranteed to work as well next time. Governments seem surprised when each recovery soaks up fewer of the unemployed. They have not taken account of the paradox of organisation, and the fact that organisations belatedly have realised that it is possible to grow without growing the labour force. Governments have been used to thinking of organisations as the delivery vehicles for social policy, not noticing that they now employ only 55 per cent of the workforce on a full-time basis, which is, in turn, only 38 per cent of all adults of working age. The organising organisation is very different from the employment organisation. Governments need to reframe their view of the world.

Paradox also confuses because we are asked to live with contradictions and with simultaneous opposites. Work will be priced highly or at zero. It looks neat to say that we should price it all in the middle, so that everyone has the same, but it would be wrong, it would not work. We have to learn to live with opposites.

To live with simultaneous opposites is, at first glance, a recipe for indecision at best, schizophrenia at worst. It need not be. My mother-in-law was generous to a fault *and* tight-fisted – she would have called it thrifty. We all knew her ways and understood. We ourselves can, in the same hour, make plans to move house next year and decide on the menu for tonight's dinner. Parents are simultaneously tough and strict on their children *and* tender and relaxed. If they do it right, the kids understand. Organisations, similarly, are tight *and* loose; concerned only about the longer term in some areas but passionate about detail in others. We all encounter and handle paradox in our daily lives. When we are used to it, and understand it, paradox is no bother. The nine paradoxes may be new, or newly important, but paradox itself has been with us forever.

It is, however, the understanding which is the key. Balancing the opposites, or switching between them, must not be a random or haphazard act. Without a clear rationale for what is happening, the balancing and the switching can be bewildering to those on the receiving end and frustrating for anyone doing the balancing. Without understanding, things do not work out as they should. Living with paradox is like riding a see-saw. If you know how the process works, and if the person at the other end also knows, then the ride can be exhilarating. If, however, your opposite number does not understand, or wilfully upsets the pattern, you can receive a very uncomfortable and unexpected shock. Children know the game well, and can take a malicious delight in upsetting the mutual understanding.

As it is with see-saws, so it is with life. If we know how and why things work, we can live with the ups and the downs, knowing that the opposites are necessary to each other. We can even come to recognise that for the see-saw to work effectively, others must get as good as we get. What follows, in this part of the book, are three general principles for living with simultaneous opposites. You could call them rules for riding see-saws. They are followed, in Part Three, by examples of the principles at work, in our organisations and in society.

3 The Sigmoid Curve

The Road to Davy's Bar

The Wicklow Mountains lie just outside Dublin in Ireland. It is an area of wild beauty, a place to which, as an Irishman born near there, I return as often as I may. It is still a bare and lonely place, with unmarked roads, and I still get lost. Once, I stopped and asked the way. 'Sure, it's easy,' the local replied. 'Just keep going the way you are, straight ahead, and after a while you'll cross a small bridge with Davy's Bar on the far side, you can't miss it!' 'Yes, I've got that,' I said, 'straight on to Davy's Bar.' 'That's right. Well, half a mile before you get there, turn to your right up the hill.'

It seemed so logical that I thanked him and drove off. By the time I realised that the logic made no sense he had disappeared. As I made my way down to Davy's Bar wondering which of the roads to the right to take, I reflected that he had just given me a vivid example of paradox, perhaps even the paradox of our times: by the time you know where you ought to go, it's too late to go there; or, more dramatically, if you keep on going the way you are, you will miss the road to the future.

Because, like my Irishman, it is easy to explain things looking backwards, we think we can then predict them forwards. It doesn't work, as many economists know to their cost. The world keeps changing. It is one of the paradoxes of success that the things and the ways which got you where you are, are seldom the things to keep you there. If you think that they are, and that you know the way to the future because it is a continuation of where you've come from, you may well end up in Davy's Bar, with nothing left but a chance to drown your sorrows and reminisce about times past.

Although he knew it not, my Irish friend had also introduced me to the Sigmoid Curve, the curve which explains so many of our present discontents and confusions. It is this curve, and what follows from it, which is the first of the Pathways through Paradox, the first of the three devices for finding a balance between the contradictions.

The Sigmoid Curve

The Sigmoid Curve is the S-shaped curve which has intrigued people since time began.

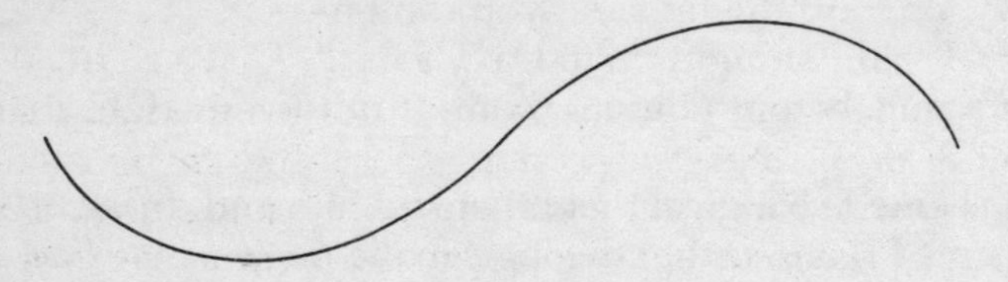

The Sigmoid Curve sums up the story of life itself. We start slowly, experimentally and falteringly, we wax and then we wane. It is the story of the British Empire – and of the Russian Empire and of all empires always. It is the story of a product's life-cycle and of many a corporation's rise and fall. It even describes the course of love and of relationships. If that were all, it would be a depressing image, with nothing to discuss except to decide where precisely on the curve one is now, and what units of time should go on the scale at the bottom. Those units of time are also getting depressingly small. They used to be decades, perhaps even generations. Now they are years, sometimes months. The accelerating pace of change shrinks every Sigmoid Curve.

Luckily, there is life beyond the curve. The secret of constant growth is to start a new Sigmoid Curve before the first one peters out. The right place to start that second curve is at point **A**, where there is the time, as well as the resources and the energy, to get the new curve through its initial explorations and flounderings before the first curve begins to dip downwards.

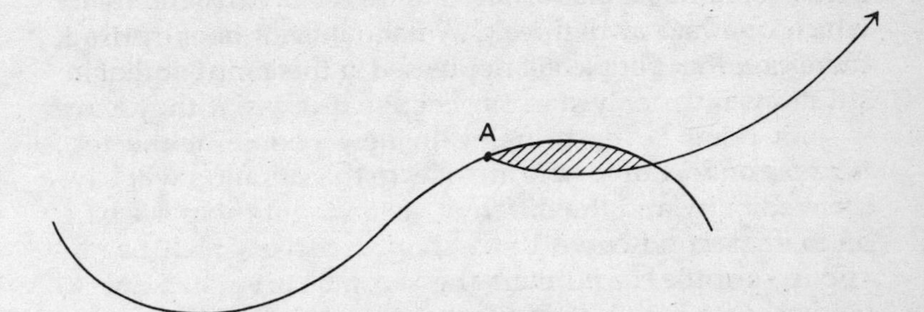

That would seem obvious; were it not for the fact that at point **A** all the messages coming through to the individual or the institution are that everything is going fine, that it would be folly to change when the current recipes are working so well. All that we know of change, be it personal change or change in organisations, tells us that the real energy for change only comes when you are looking disaster in the face, at point **B** on the first curve.

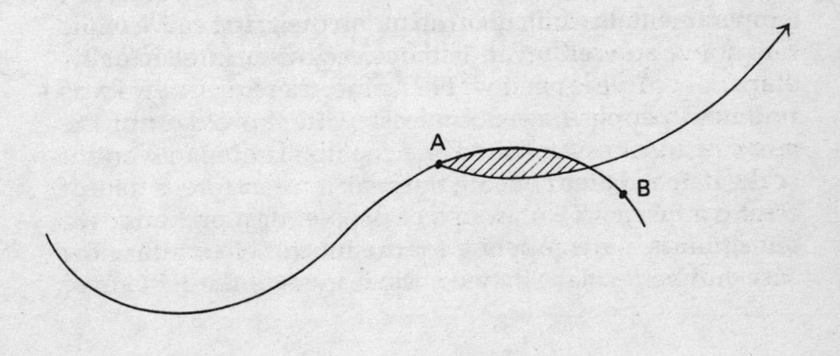

At this point, however, it is going to require a mighty effort to drag oneself up to where, by now, one should be on the second curve. To make it worse, the current leaders are now discredited because they are seen to have led the organisation down the hill, resources are depleted and energies are low. For an individual, an event like redundancy typically takes place at point **B**. It is hard, at that point, to mobilise the resources or to restore the credibility which one had at the peak. We should not be surprised, therefore, that people get depressed at this point or that institutions invariably start the change process, if they leave it until point **B**, by bringing in new people at the top, because only people who are new to the situation will have the credibility and the different vision to lift the place back on to the second curve.

Wise are they who start the second curve at point **A**, because that is the Pathway through Paradox, the way to build a new future while maintaining the present. Even then, however, the problems do not end. The second curve, be it a new product, a new way of operating, a new strategy or a new culture, is going to be noticeably different from the old. It has to be. The people also have to be different. Those who lead the second curve are not going to be the people who led the first curve. For one thing, the continuing responsibility of those original leaders is to keep that first curve going long enough to support the early stages of the second curve. For another, they will find it temperamentally difficult to abandon their first curve while it is doing so well, even if they recognise, intellectually, that a new curve is needed. For a time, therefore, new ideas and new people have to coexist with the old until the second curve is established and the first begins to wane.

The hatched area beneath the peak is, therefore, a time of great confusion. Two groups of people, or more, and two sets of ideas are competing for the future. No matter how wise and benevolent they be, the leaders of the first curve must worry about their own futures when their curve begins to die. It requires great foresight, and even greater magnanimity, to foster others and plan one's own departure or demise. Those who can do it, however, will ensure the renewal and the continued growth of their organisation.

I cannot pretend that is easy even with that foresight. I have sat and watched the chairman of a great company speak to his assembled barons. 'I have two messages for you today,' he said. 'First, I want to remind you that we are a very successful business, perhaps more successful today than we have ever been. Secondly, I must tell you that if we want to continue to be successful we shall have to change, fundamentally, the way we are working now.' He went on to explain why the different futures he foresaw would require different responses, but no one was listening. The first message had drowned out the second. If they were so successful, they felt, it would be folly to change. He was right; he was standing at point **A** and looking over the hill, but he could not get his changes implemented. Three years later, by now at point **B**, the company knew it had to change but the first person they turned on, and removed, was the chairman. He was no longer credible, nor had his conviction that he was right endeared him to his colleagues.

What is true of organisations is as true of individuals and their relationships. A good life is probably a succession of second curves, started before the first curve fades. Lives and priorities change as one grows up and older. Every relationship will sometime need its second curve. Too often, couples cling on to their old habits and contracts for too long. By the time they realise their need of that second curve they are already at point **B**. It is too late to do it together. They find other partners. On the other hand, I sometimes like to say, teasingly, that I am on my second marriage – but with the same partner, which makes it less expensive. Because we managed, in time, to find that very different second curve – together. I would not deny, however, that the hatched period beneath the peak was difficult, as we struggled to keep what was best in our past while we experimented with the new.

Capitalism, newly triumphant, probably has to reinvent itself. Things which we took for granted, like nation states and large organisations, seem to be impediments to progress not its helpers. When both monarchy and the judiciary, in Britain, are seen to be wanting, few institutions in that country can be sure that they are still on the upward curve. We ask our politicians for a lead, by which we mean a sight of the second curve, but we want them, all the same, to do nothing to disturb the first curve. In our own lives we sense that there is often another hill to climb now that life is longer and, in many ways, larger, but we have no sense of where to find it. We are, so many of us, living in that hatched area, worrying that the first curve will turn down before we find the second.

The second curve is the road up the hill to the right. We stand today at the crossroads, asking the way to our future. Words like hierarchy, loyalty and duty, no longer carry the weight they once did. Other words like freedom, choice and rights turn out to be more complicated than they seemed. What was once obvious, like the necessity of economic growth, is now hedged around with qualifications. We thought we knew how to run organisations, but the organisations of today bear no resemblance to the ones we knew, and so we have to think again, to find the second curve of management before it is too late. Meanwhile, we have to keep the first curve going. As long as we can do that, we will keep the balance between the present and the future; we can manage to live with paradox because we understand what is happening.

The Discipline of the Second Curve

The concept of the Sigmoid Curve has, I find, helped many people and many institutions to understand their current confusions. The question which they always ask, however, is 'How do we know where we are on the first curve?' One way of answering that is to ask them to make their own private and personal assessment of their position, or that of their organisation, to draw the first curve as they see it, and to mark an X on it to show where they are now. Almost invariably, when they reveal their perceptions of the curve, there is a consensus that they are farther along the curve that any of them would previously have admitted. They are nearer to point **B** than to point **A**.

Like the story of the road to Davy's Bar, you will only know for sure where you are on the curve when you look back. It is easier, too, to see where others are on their curves than to see yourself. We must therefore proceed by guess and assumption. There is no science for this sort of thing.

The discipline of the second curve requires that you always assume that you are near the peak of the first curve, at point **A**, and should therefore be starting to prepare a second curve. Organisations should assume that their present strategies will need to be replaced within two or three years and that their product life-cycles are shorter than they were. Richard Foster of McKinsey studied 208 companies over 18 years in order to discover those who were consistently successful. There were only three who lasted the course for the whole 18 years. Fifty-three per cent could not maintain their record for more than two years. Individuals should also work on the assumption that life will not continue as it has for ever and that a new direction will be needed in two or three years.

It may well be that the assumption turns out to be wrong, that the present trends can be prolonged much longer, and that the first curve was really only in its infancy. Nothing has been lost. Only the exploratory phase of the second curve has been done. No major commitments will have been undertaken until the second curve overtakes the first, which will never happen as long as the first curve is still on the rise. Keeping the two curves going will become a habit.

The discipline of devising that second curve will, however, have had its effect. It will have forced one to challenge the assumptions underlying the first curve and to devise some possible alternatives. It is tempting to think that the world has always been arranged the way it is and to delude ourselves that nothing will ever change. The discipline of the second curve keeps one sceptical, curious and inventive – attitudes essential in a time of change, and the best way of coping with the contradictions which accompany such a time.

The discipline of the second curve follows the traditional four-stage cycle of discovery. Questions start it off. The questions spark off ideas, possibilities, hypotheses. The best of these must then be tested out, tentatively and experimentally. Finally, the results of the experiments are reviewed. The first two stages cost nothing except the time for imagination. They can be very stimulating, particularly if they start from the greenfield hypothesis – 'If we did not exist would we reinvent ourselves and, if so, what would we look like?' Or, in a more personal example of second-curve thinking, 'If we did not live here, or do what we are doing, what would we be doing, where and how would we be living if we had the chance to start again?' The discipline of the second curve requires that you do not reinvent the same life, because that would merely perpetuate the first curve. The second curve is always different, although it builds on the first and grows out of it.

In *The Paradox of Success*, his book on the personal renewal of leaders, John O'Neil uses the model of the second curve to describe how leaders do, or do not, move on in life.

He points out that one essential is to let go of your past. If one is too emotionally attached to what has gone before, it is difficult to be different in any way. One can then cling on until it is too late. He quotes Odysseus as an example of a young warrior chief who was so committed to roaming and raiding, at which he once excelled, that he spent 20 years coming back from the war in Troy to his kingdom of Ithaca, reluctant to assume the responsibilities of government. By the time he did get home he was a failed commander, in rags, with his kingdom in a mess. It is the story of the man who did not want to grow up.

If success comes early, it can be particularly hard to turn one's back on it when one's star begins to wane. It was sad to watch Bjorn Borg return to the tennis courts in an attempt to recapture past glories, long after his talents had faded. It is often easier to move on from disasters than from successes. I have always, therefore, been impressed by people like Leonard Cheshire, the distinguished and heroic British fighter pilot who, after the war was over, left all that behind and set out to create a network of homes for the elderly and disabled. I am impressed by the family business in France which, at just the right time, turned its back on the textile industry in which it had made its name, and launched a chain of supermarkets. 'Where did you find the courage to do something so completely different?' I asked. 'It would have required more courage to do nothing,' the head of the family replied. 'We had the responsibility to provide a future for the family, and the past, distinguished though it was, could not have been that future.'

Curvilinear Logic

Moving on requires a belief in what Schumacher used to call curvilinear logic, the conviction that the world and everything in it really is a Sigmoid Curve, that everything has its ups and then its downs, and that nothing lasts for ever or was there for ever. Just-In-Time Manufacturing was developed in Japan, and later copied everywhere. The idea of a constant stream of deliveries to your factory door, as and when you needed them, was blindingly obvious when you thought about it. Cut out the warehouse and all those storage costs. Let the suppliers carry the inventory costs instead, or rather, eliminate them completely, provided always that you can guarantee that the lorries with the bits will arrive 'just-in-time'. Unfortunately, the idea became too popular. They tell me that the delivery vehicles now jam all the freeways around Tokyo, meaning that just-in-time often gives way to just-too-late. The costs of the traffic jams are beginning to outweigh the costs of the original warehouses, to say nothing of all the environmental damage caused by those idling exhausts. You can have too much of a good thing, or, curvilinear logic strikes again.

Curvilinear logic is not intuitively obvious if you are still ascending the first curve. Business history is littered with the stories of founding fathers who thought that their way was the only way. The French textile business mentioned above is a notable rarity among family businesses. The paradox of success, that what got you where you are won't keep you where you are, is a hard lesson to learn. Curvilinear logic means starting life over again, something which gets harder as one gets older. It is better, therefore, in organisations, to entrust the curvilinear thinking to the next generation. They can see more clearly where the first curve is heading and what the next curve might look like. It is the job of their elders to give them permission to be different, and then, when the next curve is established, to get out of the way. For that to happen, there has to be a new curve for them, outside.

'My father brought me back from America to run the business here in Treviso,' his daughter said. 'But he still comes into the office every day, even Sundays. He wants me to run the business as if I was him, and I'm not. And the business has to change, if he would only let it. It's very frustrating.' Her story was not unusual. The father had nothing else he wanted to do. The business had been his life, and now he had no other. 'Wet leaves, we call them in Japan,' said the Japanese lady, describing the reaction of Japan's women to their retired executive husbands. 'You know how it is with wet leaves, they just stick around!' For curvilinear logic to work in the organisation there has to be a life beyond the organisation for the heroes of the first curve.

The Coca-Cola Company is, on the face of it, the great exception to the concept of the second curve. For 104 years they have sold the same product in the same packaging with much the same advertising. The only time they changed the formula they were forced by their customers to reverse the decision. Their secret may lie, however, in the motto which is inscribed in their central offices and in the minds of all its officers – 'The world belongs to the discontented'. It was the favourite saying of their early and long-time chairman, Robert Woodruff. He was warning against complacency and advocating a perpetual curiosity – the itch of the second curve. Coca-Cola's Japanese company, I was told, test-markets a new soft-drink variety or other product every month. Even if most of them fail most of the time, it keeps the questing spirit alive. When and if Coca-Cola's 104-year curve turns down, they hope that they will be prepared.

The Japanese, of course, have their own word for it – *kaizen*, or continuous improvement. The assumption behind *kaizen* is the assumption behind this book – that there is no perfect answer in a changing world. We must therefore be forever searching. Anita Roddick, of the Body Shop, puts it more succinctly: 'What is so wonderful about the Body Shop is that we still don't know the rules.' As long as they think that way, so long will they thrive. Complacency is the enemy of curiosity.

The Royal Dutch Shell Group has yet another approach. They call it scenario planning. It has been well explained by Peter Schwartz, one of the early members of the planning group, in his book about it, *The Art of the Long View*. A group of executives, aided by some outsiders, spends a year or more drawing up alternative scenarios for the oil business and the countries and cultures in which it operates. These are not plans but possibilities, deliberately set at the opposite ends of a spectrum. The planning group then uses these scenarios educationally, exposing their managers around the world to the alternatives and asking them to consider how they would respond if either happened. Shell want no surprises, and were not surprised by the oil crisis of the early 1970s nor by the collapse of the Russian Empire. Their second-curve thinking was ready. It was not so, says Schwartz, in the case of the American military. They made every sort of contingency plan for the Cold War but they never asked the scenario question 'What if we won?' When they did win, they knew not what to do.

Peter Senge, in his classic book on the learning organisation, reminds us that our mental models, or private scenarios, are crucial to the learning process. We all carry mental maps around with us – that hierarchy is natural, for instance, that women can't manage or that men don't care; that careers last until 65 or that every next job has to be a promotion. We need to check that these assumptions are still valid because they lock us into our existing curve. They inhibit second-curve thinking. My first book on organisations was written 20 years ago. Quite unconsciously, I used the male pronoun exclusively throughout the book. It became a standard textbook, used by those training to work in schools, hospitals and social services as well as business. My book caused a great deal of offence to the many women who had to study it because it appeared to imply that I, the supposed authority, thought that there was no place for them in management. My unconscious mental map of 20 years ago only mirrored what many men felt then, and some still do. That map locked them into their first curve; it made it difficult for them to envisage another kind of world and another way of doing things. My book was not only offensive, it was harmful.

Many of the ideas in this book stem from second-curve thinking – the discipline which says that the past might not be the best guide to the future, that there can be another way, and that some 'myths of the future', as Schwartz calls them, will help. We must, however, beware that we do not abandon the first curve too early. The second curve needs the resources and the time which only the first curve can provide. It has to grow out of the first. 'Dreams give wings to fools,' my young daughter used to tell me when she heard me fantasising about other lives which we might live. She was expressing her instinct that the future needs to be rooted in the past if it is to be real. The secret of balance in a time of paradox is to allow the past and the future to coexist in the present.

Fertilising the Second Curve

Second-curve thinking will come most naturally from the second generation, those who will inherit the future of the institution or the society. They will, however, need both permission and encouragement. They must realise that what they might privately think of as revolution, or even sedition, is possibly the way ahead in due course. New ideas can coexist with old.

One organisation openly entrusted its second-curve thinking to a group of executives in their early thirties. It happened, however, almost by accident. They wanted to celebrate the twenty-fifth anniversary of their organisation. Their first thought was to commission a history of those first 25 years. That seemed, on reflection, to be too self-indulgent and uninteresting. They therefore decided to commission an outside look at the next 25 years for their industry. They were persuaded that the most fruitful way to do that would be to entrust that forward look to the brightest and best of their own people, people who might be leading their organisation when those years arrived. The look at the future should, therefore, include some thoughts and recommendations on how the organisation should adapt to the changes they might foresee for their industry and the world around them. They were giving these young people the responsibility for their inheritance.

I was asked to act as mentor to the study. I agreed, provided that the board of the organisation agreed to publish the non-confidential part of the exercise as a booklet, without censoring it in any way. The board agreed, but went further. They offered to invite all their customers to a reception to celebrate the anniversary, to listen to a presentation of the findings of the group, whatever they might be, and to receive a free copy of the uncensored booklet. The effect of this advance commitment was impressive. The group saw that this was not some ingenious educational exercise but a genuine attempt to build some new thinking into the existing fabric of the organisation. They were begin publicly trusted by their seniors to develop some new thinking. Their seniors were not only impressed by the results of the study, they took them to heart. Their advance commitment had ensured that they would not feel it necessary to defend the status quo, the first curve, and squash the beginnings of the second.

It is important that the seniors give permission and encouragement. It is also important that the next generation accepts their responsibility for second-curve thinking. Preoccupied with the immediacy of their own careers, it is tempting to think that second-curve thinking can be left until later, that the present is their proper priority, the future the priority of those in charge. In actual fact, it should be the other way round.

I helped, once, to organise what came to be called the Windsor Meetings. They took place at St George's House, a small study centre in the middle of Windsor Castle, often used for weekend gatherings of influential individuals to discuss, privately and informally, social and ethical issues. They were, inevitably, discussions about the present, because the people who came were in charge of the present. We decided, with the help and support of some businesses, to bring together representatives of the next generation of influential people, individuals from all sectors of society who were identified for us as likely future leaders in their spheres.

Thus it was that a young colonel, tipped as a future general, found himself beside a rising trades-union official, a talented young headmistress, a banker, some civil servants, three of the younger and more thoughtful politicians from the different parties, a campaigner for human rights, the new editor of one of the quality newspapers, a television newscaster, a doctor and a lawyer, five business executives – all people successful in their thirties, but preoccupied with their own careers, too busy, at this stage, to look outside or to know anyone not involved in their line of work. They were all at the lower end of their personal curves, and rising fast. Invited to Windsor Castle, as guests, for a week, they were asked to debate and discuss the shape of the society which they would inherit, as people likely to reach the higher level of their spheres of influence.

Few of them had thought about such broad issues. None of them had met such a wide range of other interests. Their discussions were always stimulating and their reports insightful, but the ultimate benefit was the realisation that they had a responsibility to help shape the society which they were likely to inherit. It was a consciously élitist exercise, because if those soon to be in power are not conscious of their responsibility to shape a second curve, who is?

Many of those groups still meet, because, in spite of their very different preoccupations, they found that they also shared a concern for the future of their society, that it should be civilised as well as rich, humane as well as adventurous. There is strength in companionship when it comes to shaping the second curve. We have to hope that, when and if they reach positions of eminence, they will not forget their commitment to that second curve.

Both of these examples used insiders. Some organisations prefer outsiders, feeling that they may have a more objective view and a clearer perspective. Consultants thrive on contracts for what are, effectively, second-curve thinking. The thinking, however, is only part of it. There needs to be the commitment to carry it through, to endure the early dip before the curve climbs upwards, to live with the first curve while the second one develops. These things cannot be done by outsiders. To manage paradox, you need to live with it as well as analyse it.

4 The Doughnut Principle

The Inside-Out Doughnut

The doughnut in question is an American doughnut, the kind with a hole in the middle, rather than the British version, which has jam instead of a hole. The doughnut principle, however, requires an inside-out doughnut, one with the hole on the outside and the dough in the middle. It can only, therefore, be an imaginary doughnut, a conceptual doughnut, one for thinking with, not eating.

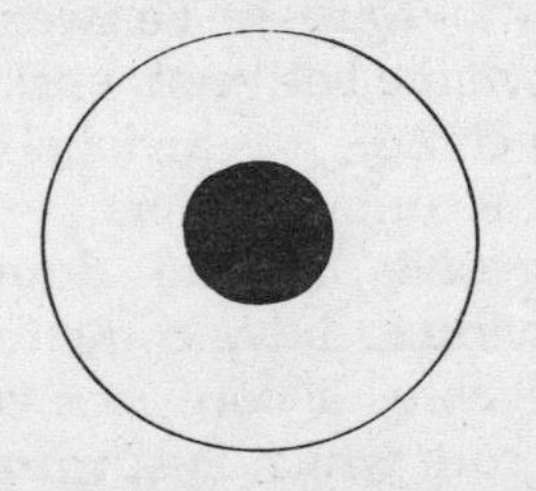

A doughnut may seem to be an unlikely Pathway through Paradox, but the concept of balancing a core and a bounded space is crucial to a proper understanding of most of life, as I shall hope to show. It is a way to find the balance between what we have to do and what we could do or could be. It is a way of getting around the problem of the empty raincoat, of being an instrument of society but also a free individual.

We might look, for instance, at our job, whether that is our paid job or our unpaid role in life, as parent, wife, husband or carer, student or friend. The heart of the doughnut, the core, contains all the things which have to be done in that job or role if you are not to fail. In any formal job, these things will be listed and will be called your duties. Even where they are not listed, these duties are often well understood. The core, however, is not the whole of the doughnut. If it were, life would be all chore as well as core. There is, thankfully, the space beyond. This space is our opportunity to make a difference, to go beyond the bounds of duty, to live up to our full potential. That remains our ultimate responsibility in life, a responsibility which is always larger than our duty, just as the doughnut is larger than its core.

The doughnut image is a conceptual way of relating duty to a fuller responsibility in every institution or group in society. Doughnuts stimulate our thinking about the proper equation between commitments and flexibility, in all the structures of our work as well as in our personal life. We can draw a doughnut to represent a relationship, or an organisation, or a work group, just as we can use it to reveal the balance in our own life between work and family or between necessity and choice. It is a visual tool for balancing what often seem to be contradictions.

Much of life now looks like that doughnut. Organisations as well as individuals have come to realise that they have their essential core, a core of necessary jobs and necessary people, a core which is surrounded by an open flexible space which they fill with flexible workers and flexible supply contracts. The strategic issue for organisations, nowadays, is to decide what activities and which people to put in which space. It is not always obvious. Businesses have their core obligations or duty to their shareholders, but their responsibilities go much farther. Finding the right balance between this duty and a wider responsibility is a dilemma at the heart of capitalism.

Schools, in most countries, now have their required core curriculum, with discretionary space around it. The argument, again, is about balance. What, and how much, should go in which space? Too much core and there is no room for individual student difference or for local school initiatives. Too much space and there is too much variety in the standards of delivery.

We can apply the doughnut principle to processes as well as to structures. Reward systems tend to lay down the minimum remuneration with space for options, bonuses and performance money beyond. Our implied contracts within our personal relationships contain a core of obligations, with space for individual difference around the core. In your marriage, wrote Khalil Gibran in his poem, *The Prophet*, which gets read at countless weddings and is then, too often, forgotten, you should stand together, but 'let there be space in your togetherness'. What every couple needs to define, however, is what should go into that space and what the boundaries are to be. A marriage without boundaries is, I suspect, an unreal doughnut, and doomed to fail. Every couple could, with advantage, draw their own doughnut.

The doughnut principle starts early. I remember my own schooldays. I had passed the big examination and was studying the certificate when I was confronted by my disappointed teacher, obviously quite distressed by my results. 'What's wrong?' I said, 'I passed, didn't I? Isn't that enough?' 'Enough is never enough,' he replied, 'until you have exhausted your potential. To pass was easy. You could have done better, much better.' Enough is never enough. I recalled John Donne's lines, 'When you have done, you have not done, for there is more.' To pass was the essential core. To fill the doughnut I had to do more. Life, my teacher was trying to tell me, should not be a half-filled doughnut. I have spent most of the rest of my life wondering how I ought to fill it.

That principle now applies to much of our work. That ugly word 'empowerment' could better be interpreted as the doughnut principle at work. In the past, jobs used to be all core, certainly at the lower levels, because too much discretion meant too much unpredictability. One of my early jobs had the fine-sounding title of Regional Co-ordinator Marketing (Oil) Mediterranean Region. My friends were impressed, but they did not know the reality. The reality was a three-page job description outlining my duties, but the hard truth was contained in the final paragraph: 'Authority to initiate expenditure up to a maximum of £10'. My doughnut was all core and no space. That way the organisation got no surprises, or so it hoped. All was predictable, planned and controlled. It was also dull and frustrating, with no space for self-expression, no space to make a difference, no empowerment. My memoranda went from my role – MK/32 – not from me. I was merely a 'temporary role occupant' and I felt like an empty raincoat.

There are some, on the other hand, whose jobs are nearly all space, with little core and no boundary. Ministers of religion have a visible core to their work – the church services, some sick visits, committees and finances – but there is no limit to their responsibilities for the souls of their congregation or for their evangelising work. Some of the most stressed people that I have known have been people with jobs like these, because there is no end, no way in which you can look back and say 'It was a great year', because it could always have been greater. Empowerment, in a sense, has gone too far. Without a boundary it is easy to be oppressed by guilt, for enough is never enough. Entrepreneurs revel in the space for discretion, but the successful ones are careful to give themselves targets and limits. Even then, the record of early entrepreneurship is one of long days and no holidays, of unremitting effort to fill the space beyond the core. A sensible job is a balanced doughnut.

Type 2 Accountability

More space, in fact, is not always welcomed, even in a conventional job. More space means more choice but also, paradoxically, more room for error, or, more precisely, another type of error. In statistics, as I recall, there are two types of error. There is a Type 1 error which, in simple terms, means getting it wrong, and a Type 2 error which, in effect, means not getting it right, or as right as it could have been. There is an important difference. A Type 2 error means that the full possibilities of the situation have not been exploited or developed; enough was not enough. In the old tightly planned world in which everything was contained in the inner circle, you only had to look out for the Type 1 errors. As long as you avoided those you could call yourself successful.

Management was easier, too, since the priority was to check for Type 1 errors. Avoid those and the system was designed to deliver. For many people, life was keeping one's nose clean, compiling, as the years went by, a fault-free curriculum vitae, lived by the book, with retirement as the promised rest at the end, just as the life-insurance advertisements would have it. 'He walked in many corridors of power,' someone said of a politician, 'and left no footprint in any of them.' It was a life free of Type 1 errors.

It was also a life devoid of the Type 2 errors. These are the errors of omission not commission, the things one did not do which one could have, the failure to fill the space between the circles. The old prayer book of the Anglican Church puts it nicely, I discovered, as I mumbled my way through the familiar words one day: 'We have left undone those things which we ought to have done [Type 2]; and we have done those things which we ought not to have done [Type 1].' I used to think that what was meant by the first bit were those chores I had neglected, the difficult meetings I had put off, the letters I had not written, but those, I realised, are all Type 1 errors. The important sins of omission are the things I did not do, which could have made a difference. Enough is not enough.

We hanker after the freedom of more space in our lives

and our work. Leaner, flatter organisations provide that space; but now it is up to us to fill the space. We used to be held accountable only for Type 1 errors. Now we have a new accountability, for the things we could have done but didn't. The two accountabilities are a new fact of life. With space goes responsibility. Only when that is generally accepted will we be able to have a truly free society, one in which the freedom to be what you want to be is accompanied by the responsibility to do no harm to others (Type 1) and to use the freedom to some purpose (Type 2).

One day, too, our public bodies will recognise that it is not enough to make no mistakes, Type 1 accountability, it is also important to have done the work as well as it could have been done, to have been better than expected. Public accountability needs to be redefined to include the recognition of Type 2 responsibilities. John Major's idea of the Citizen's Charter in Britain is a small step towards this recognition, and will be a bigger step when it recognises that there are rewards for excellence in service, as well as penalties for lapses.

Personal Doughnuts

Some people make their work the whole of their life. That necessary core of the job fills the whole doughnut, leaving little or no space for anything else. Are they right or wise? There is an argument that capitalist businesses do not exist as, theoretically at least, communist structures did, to liberate and develop people's humanity and allow them to become moral and fulfilled human beings. Existential development, says Britain's Elizabeth Vallance, is not the primary aim of business but of churches or educational or artistic institutions. Businesses will look to the self-development of their people only to enhance the business's ability to make profits. If she is right, then those who seek their fulfilment in demanding business jobs are likely to be disappointed. It would probably be no better if they worked full-time for her churches or educational or artistic establishments. In all the organisations of a capitalist society, the individual is, in strict theory, the instrument not the purpose.

Against that, there is the view that all work should be a calling or vocation; that the wealth creation of business is as worth doing and as valuable as the health creation of a hospital. We can and should, that argument goes, get our fulfilment out of our work. There can be no single answer. The doughnut principle would suggest that if you cannot get your existential development from your current job you should either change the job or make sure that the empty spaces in your personal doughnut are filled somewhere else. One job does not have to fill all needs.

I used to think that it should. I looked for a job which would provide me with interesting and exciting work; work, too, that I would be proud of doing. I also wanted enough money and the chance to make more of it, if I needed to, good companions and a pleasing location with the chance of travel. Needless to say, I never found the perfect job. There was, however, a doughnut solution. If I adopted a 'portfolio' approach to life, meaning that I saw my life as a collection of different groups and activities, of bits and pieces of work, like a share portfolio, I could get different things from different bits. A part of that portfolio would be 'core', providing the essential wherewithal for life, but it would be balanced by work done purely for interest or for a cause, or because it would stretch me personally, or simply because it was fascinating or fun.

It was easier, I found, to make money, if that was all that you were bothered about, than if you tried to combine money-making with all the other attributes. Similarly, it was easier to find work that was involving and worthwhile if you weren't too concerned about the pay. It meant, however, turning down the offer of one of the 70-hour-a-week jobs which would have left no time for anything else, and putting together a package of different kinds of work, a work portfolio. My life, now is doughnut-shaped. I can even specify the amount of days which I am prepared to allocate to core activities and the amount left over for personal space. As I get older, the core begins to shrink, leaving me with the interesting but difficult problem of how best to fill the space beyond the core, to live up to my responsibility for my life.

Increasingly, it should be possible to arrange a portfolio of different sorts of work within the same organisation, by joining a number of their different doughnuts. Wise organisations recognise the advantages of these internal portfolios. Different tasks and different groups bring out different talents in the individual; they confront him or her with different experiences. Some businesses now actively encourage their staff to take on some voluntary activities in the community, allowing them time from work if necessary. Other organisations are happy to see their executives sitting on public bodies, teaching a course at a local college, serving on a school board or standing for political office. It is, they say, an excellent form of development. It is also a way of building a work portfolio with the company's blessing.

There are fewer, smaller cores in the job doughnuts these days, in all spheres and levels of work. If we wait around for someone to tell us what to do, we shall wait for a long time. If we look for a standard route through life, a sure way, guaranteed to get us through to the end, we shall be disappointed. We have to fill our own spaces.

Societies which overemphasise the core can be too regimented. A place for everyone and everyone in their place was Plato's version of a just society, but it meant that everyone's role in life was predetermined, all core and very little space for individuality. It was an idea that lingered on in Britain when I first came to live and work in that country, even though the core was more socially determined than official. There were strong norms on dress and behaviour. 'Wear brown on Sundays,' I was told, 'and never telephone anyone after 10 o'clock at night.' I envy the freedom of my children and their friends who respect neither of those conventions nor many others, but they can suffer from the burden of too much space. There is too much choice of career, too many varieties of life styles from which to choose. It is no longer mandatory to be married in order to be a parent, let alone to share a home. It is hard for them to see what the core of life can be apart from earning the bare necessities, and even those will, if need be, come from the state.

They have the freedom to design their own doughnuts. They would be wise to give themselves both core and boundaries, a base-line for the kind of life they want to live and some areas they will not touch, some things they will not do, some rules of conduct which they will keep. A society which emphasises rights but neglects obligations can leave too much space for its citizens. The problem for the unemployed is not so much that they are hungry but that they have no core to their lives. Empty doughnuts are not easy to live with, any more than doughnuts which are all core.

Doughnut Organisations

Work, itself, is no longer organised as it used to be. Organisations are not now drawn as pyramids of boxes. British Steel is said once to have had an organisation chart which, when unfolded, stretched across a room. Those charts now have circles and amoeba-like blobs where the boxes used to be. It isn't even clear where the organisation begins and ends, with customers, suppliers and allied organisations linked into a varying 'network organisation'. Work no longer means, for everyone, having a 'job' with an employer. As organisations disperse and contract, more and more of us will be working for ourselves, often by ourselves.

The new shape of work will centre around small organisations, most of them in the service sector, with a small core of key people and a collection of stringers or portfolio workers in the space around the core. David Birch, an economist with the research organisation Cognetics, studied the job market in the USA from 1987–91. He found that the big firms laid off a net 2.4 million workers in that period, while firms with fewer than 20 employees added 4.4 million new jobs, with slightly larger firms adding another 1.4 million. Nor were all these jobs, any longer, hamburger jobs. Software, telecommunications, environmental engineering, health products and services, and specialised education are increasingly the province of the tiny partnership. They are all well suited to the portfolio worker, who costs much less if the firm does not have to house him or her. Their work, and that of their larger brothers, increasingly fits the doughnut pattern.

We can see the doughnut pattern most obviously when we look at the new-style organisations. The ½ × 2 × 3 formula, which all organisations have to work towards in this competitive age, means that every organisation these days has its smaller core and its surrounding partnerships. Some of these partners are their traditional suppliers, some are independent professionals, some are the part-time peripheral workforce and some are the allied businesses, partners in joint ventures of one sort or another.

The British government, in an attempt to reduce costs, is 'market-testing' many of its traditional core activities. Functions which have traditionally been done by the core Civil Service are now required to be tested for cost and performance against outside tenders. If they compare unfavourably, the activity moves into the space of the doughnut. The collection of Britain's National Insurance contributions, a form of wages' tax worth nearly £40 billion a year, is one activity proposed for such market-testing. There are, however, some reasonable concerns that it would be too risky to contract out such a vital source of national revenue. It should, some feel, be designated a core activity, impervious to market-testing. The Home Office is contemplating putting activities as diverse as the Criminal Injuries Compensation Board, the customs control at ports and airports and even the Research and Planning Unit out to tender. The BBC's contentious policy of 'producer choice', under which all decisions have to be market-tested, means that the doughnut balance of the Corporation is being decided piecemeal, by individual producers, on a short-term cost basis. Many fear that this process is not guaranteed to be the best way to arrive at a long-term strategic balance, at the optimum doughnut for the BBC.

Businesses routinely put their materials-suppliers in the space of their corporate doughnuts. The vertically integrated organisation, one which wanted to own and run the whole of its doughnut, is a thing of the past. Some, however, also contract out crucial service functions. Eastman Kodak sees sense in contracting out their whole information system. Others give their strategy formulation to consultants. There is not limit to what you can, if you wish, put in the space of the doughnut. It is the balance which is crucial. The British Civil Service is worried about the effects on morale if too many of the key elements of their work are given to outsiders. Short-term savings may result in long-term damage if a demoralised service fails to recruit new talent as a result. There is no neat general answer. It is always a question of finding the appropriate balance.

There can be worries that the new partners may effectively become part of the core if they are bonded too tight. The flexibility, which was the point of the doughnut structure, disappears if the supplier becomes dependent on one customer for most of its business, or the firm on one supplier. Contracts should be flexible, experience suggests, and not more than 30 per cent of capacity or requirements should be tied to any one outfit.

Ricardo Semler, president of Semco in Brazil, deliberately designs his whole company as a doughnut, or what he would call a double circle, with a group of counsellors in the middle and all the other workers, the partners and associates, as he terms them, in the outer space. They are then held together in smaller doughnuts, or circles, by co-ordinators. Managing doughnuts is the new organisational challenge. It is a challenge because one is managing the doughnut, and its different spaces, rather than the person. Managing other people's spaces is not easy. It is no longer the manager and the managed, but the designer of the doughnut and the occupant; a different relationship, one built more on trust and mutual respect than on control.

Organisations everywhere are being 'reinvented' or 're-engineered'. They are breaking down, or rather, blowing up their functions and their old ways of working and are regrouping people, equipment and systems around a particular task. They are creating work doughnuts, groups with complete responsibility for discharging the task, with specified rules and duties – the core – and a lot of discretionary space to do it in the way that they think best. The results can be startling. The most famous is Ford's accounts-payable department which thought that they had done well when they reduced their staff by 20 per cent to 400 until they found that Mazda did the job with only five! Ford thought again and got their numbers down to 100, the ½ × 2 × 3 formula beaten twice over. The term 'Business Re-Engineering' has been patented by a consulting firm, but the concept at its heart is as old as the doughnut.

The doughnut organisation is even laid out, physically, as a doughnut. The centre no longer dominates from a headquarters tower block. It is smaller and more club-like, with outlying or satellite offices around the country. Frank Becker, who is leading an American project called Workscape 21 at Cornell University, believes that more and more workers will split time between a central office, a computer-equipped home office and a satellite office in a suburban business park. The central office itself will be doughnut-shaped, built around 'common rooms' which will increasingly resemble hotel lobbies, or the rooms of a clubhouse.

I was meeting with the senior executive group of one of the country's leading manufacturers of office furniture. The subject for discussion was the executive suite of the future – what sort of furniture would it require. We decided to start by quizzing Anita, the Human Resources director, on how she spent a typical working week, in order to get some feel of how she would use her office. In the last month, it turned out, Anita had worked out of hotel rooms, airplanes and airport lounges, in the premises of the subsidiary businesses, and in her home in the evenings and early mornings. She had spent two Friday afternoons in her office, collecting the non-urgent mail which had not caught up with her on her travels, and checking schedules with her secretary. The only classy furniture she really needed was a bag to hold her bits and pieces of electronic equipment. For her, the office was a club which she checked into occasionally.

Doughnut Thinking

This has, for me, been the year of funerals. I have listened to many a eulogy on the life of old friends or relatives. Those eulogies always relate the facts and the distinctions of the life now ended, but they go on to the important and most interesting bit – a description of the person as they were to those who knew and loved them. By the time we die, I realised again and again, our doughnuts have to be complete. The real 'us' is, ironically, not in the core but in the whole, if we think of the core as the necessities of life, that curriculum vitae which most of us so assiduously compile in our early years. It is up to us, and only us, to fill the space before we die.

I seldom know, in any detail, what it is that my friends really do in their work, because when we meet we talk of other things. Ironically, I think, I like them better when they are not succeeding, and they me likewise, because there is then more space and time for friends and fun. To put it crudely, they, and I, are less boring when less successful. There is, then, more space in our doughnuts.

We overdo the core. In our personal lives we often exaggerate the necessities. Few need as much as they think that they do, or as much security as they hanker after. Organisations build bigger cores than they need, and impose bigger cores on their internal doughnuts than are necessary. Schools, everywhere, impose timetables on their students, filling all their hours with things they are required to do. In a world which seems crammed with rules and duties, responsibility goes unregarded. Even the rules and duties become devalued, as we instinctively reject the idea that so much of us can be prescribed. No one wants an empty doughnut, one with no obligations and no commitments, but one with too large a core breeds a sense of impotence.

We find it a paradox that people clamour for rights but ignore their responsibility, that people want democracy but expect others to sort out all the problems for them, that they complain when others take the initiative but take no initiative themselves. We find it odd that there is so little time to enjoy the fruits of our labours, but later find that we don't know how to enjoy the fruits when we do have time. We are unused to all that space in our lives. We have been so burdened with duties that we have never learnt the delights of responsibility, of making a difference to someone or something.

By overemphasising the duties and the rules of the core, organisations unintentionally breed distrust. A £10 discretionary authority hardly indicates any confidence in one's judgement or integrity. Obsessed by the need to control, organisations create self-fulfilling prophecies when their people find that the only way to be independent is to break the rules. My children learnt to smoke because the schools they went to made non-smoking a rule. They justified the rule with all the lessons about the dangers of nicotine, but the implicit message was, 'We don't believe that you will heed these warnings so we will make it a core obligation.' Such a deliberate denial of responsibility made it legitimate, in the eyes of the students, to ignore the rule. Smoking became a symbol of free choice, a personal space in their doughnut.

If we do not allow people space, we cannot expect responsible behaviour. There are risks, of course. Not everyone can handle the same amounts of space and responsibility. Doughnuts have to be adjusted to the capacity of the individual or the group. As parents we allow more space as the children grow, but always within boundaries. The risks of not doing it, the Type 2 error of restricting space, are, however, much more serious. Too much space can cause an error or an accident. Too little can impoverish a life. We were not meant to be an empty raincoat.

5 The Chinese Contract

I remember my first exposure to the 'Chinese contract'. I was the manager in South Malaysia for an oil company, responsible amongst other things for negotiating agency agreements with our Chinese dealers. I was young, enthusiastic and, I suppose, naïve. After the conclusion of one such negotiation, the dealer and I shook hands, drank the ritual cups of tea, and were, I felt, the best of friends. I took the official company agency agreement out of my case and started to fill in the figures, preparatory to signing it. 'Why are you doing that?' asked the dealer in some alarm. 'If you think that I am going to sign that you are much mistaken.' 'But I am only writing in the figures which we agreed.' 'If we agreed them, why do you want a legal document? It makes me suspect that you have got more out of this agreement than I have, and are going to use the weight of the law to enforce your terms. In my culture,' he went on, 'a good agreement is self-enforcing because both parties go away smiling and are happy to see that each of us is smiling. If one smiles and the other scowls, the agreement will not stick, lawyers or no lawyers.'

I think that I persuaded him that it was just a piece of company ritual and of no significance, but the episode sent me away thinking. I had grown up in a culture which believed that a good negotiation was one in which only one of us, myself, came away smiling, but concealed that smile lest the other guess that one had got the better of him or her. Negotiation was about winning at the expense of the other party. You then had to enforce your side of the deal, using the law or the threat of the law. I had met a culture where negotiation was about finding the best way forward for both parties. No wonder we needed so many more lawyers in our culture.

The Chinese contract, I later realised, embodied a principle which went far beyond the making of lasting commercial deals. It was about the importance of compromise as a prerequisite of progress. Both sides have to concede for both to win. It was about the need for trust and a belief in the future. Writ large, it was about sacrifice, the willingness to forego some present good to ward off future evil, or, more positively, it was about investment – spending now in order to gain later.

We have no chance of managing the paradoxes if we are not prepared to give up something, if we are not willing to bet on the future and if we cannot find it in ourselves to take a risk with people. These are our Pathways through the Paradoxes, if we have the will. The pursuit of our own short-term advantage, and the desire to win everything we can, will only perpetuate animosities, destroy alliances and partnerships, frustrate progress, and breed lawyers and the bureaucracy of enforcement.

The Chinese contract, as I discovered to my chagrin, involves a major rethink of our cultural habits, even in China, where they may not appreciate my magnification of their trading habits into a principle of life. The pursuit of self-interest has to be balanced, as Adam Smith's two books remind us, by 'sympathy', a fellow feeling for others which is, he argues, the real basis of moral behaviour. Only if we are conditioned by this 'sympathy' will we want to take any risks with our fellow men and women, will we trust them farther than we can count them, or want to make life better for those we never meet. As Arthur Okun put it: the 'invisible hand' needs to be accompanied by an 'invisible handshake'. Self-interest, unbalanced, can only lead to a jungle in which any victory will mean destroying those on whom our own survival will ultimately depend. That would be the paradox to end all paradoxes. The tragedy of the commons, it was labelled, when the individual farmers maximised their own short-term use of the common land only to find that, when everyone did the same, the land deteriorated until all the grazing failed.

There are those who think that 'sympathy' will remain a very weak force, always yielding to self-interest. The evidence, however, is against them. Jean Piaget studied young children playing, and observed an inherent sense of fairness, particularly in older children who had longer time horizons. 'It makes her happy' or 'I don't want to see him cry' were the common explanations for gifts of generosity, and there was more generosity than there was hoarding. The Chinese will be relieved. Their policy of one child per family means that all young children spend their early years without siblings. The Chinese have, therefore, started courses in 'sharing' in the primary schools – to help them learn the principle of the Chinese contract! Adults don't always lose the habit. Most people don't put up the price of candles in a power strike or shovels in a snowstorm. There is sympathy in humankind as well as greed and cruelty.

The Morality of Compromise

The 'morality of compromise' sounds contradictory. Compromise is usually a sign of weakness, or an admission of defeat. Strong men don't compromise, it is said, and principles should never be compromised. I shall argue that strong men, conversely, know when to compromise and that all principles can be compromised to serve a greater principle. I have said 'strong men' because I am assured that strong women have always known the value of compromise in the interests of progress.

Most of the dilemmas which we face in this time of confusion are not the straightforward ones of choosing between right and wrong, where compromise would, indeed, be weakness, but the much more complicated dilemmas of right and right. I want to spend more time on my work, *and* with my family; we want to be good corporate citizens *and* return a decent profit; we want to trust our subordinates *but* we need to know what they are doing. At other times, the interests of different parties are in conflict. Without compromise by both, there will be no movement. Stalled by a refusal to make concessions, things stagnate. Progress is sacrificed to pride.

I once listened to Lord Owen, then Dr David Owen. He had just finished his stint as the Foreign Secretary in Britain's Labour government of the Seventies and was not yet the leader of the Social Democrats. He was addressing a group of bishops on a topic of his own choosing – the morality of compromise. He had, he said, one August, while on his own in the Foreign Office, when most of the staff were on holiday, received a request to provide riot and control equipment to the Shah of Iran, whose regime was fast falling apart. The Shah, he said, and his regime were deeply repugnant to him; they offended all the principles of democracy and social justice in which he and his party believed. Nevertheless, as he weighed the alternatives, with no one around to give advice, he had concluded that to meet the request was preferable to the only alternative, a shoot-out by the Shah's army with bodies dead in the street. He gave support to a regime he abhorred because the alternative was worse. One principle was sacrificed to a greater principle.

If you stick too fast to your position, he told the bishops, you may have the comfort of feeling that you are in the right, but you may also have committed the greater wrong of stalling any movement in what is generally the right direction. It is ironic that, some years later, it was David

Owen's too-rigid adherence to his own position which helped precipitate the demise of the party he led. Years later again, he found himself trying to persuade Serbians, Bosnians and Croatians to compromise in order to move forward, towards peace in the old Yugoslavia.

Peter and Pam Richards grow fruit and vegetables on their farm in the Channel Islands. The market is very competitive, they say, and life is not easy. They are also signed-up members of the Friends of the Earth, and ecological enthusiasts. 'But,' says Peter, 'if we did not cover our new potatoes with polythene they would be ruined in two weeks.' 'We have had to learn to balance our idealism with pragmatism,' said Pam, 'to compromise. I suppose,' she added, 'that it's part of growing up!'

There is, however, a seductive power in certainty. If you have no doubt, then you never see the need to compromise. Unprincipled people are not, in fact, unprincipled. They have, instead, one overriding principle, which may be self-interest, or the good of the business or the nation as they see it, or even what they might call 'the will of God'. These principles they will not compromise. There will be no way forward with them, except on their chosen road and in their chosen vehicle. For them, compromise is a weasel word, signifying capitulation. Their certainty gives them power, but at a cost. There is no room for anyone else in their camp, except for converts. Such people never see the need for a second curve, and, in time, their curve turns down. The certainty and self-assurance of a good leader must be tempered with the spirit of compromise if others are not to feel excluded. Margaret Thatcher was one who never felt the need to compromise. Her certainty gave her strength, and was much admired, even by those who disapproved of what she did with it. But in the end, the refusal to compromise brought her down. Compromise is essential to democracy. It is also essential to leaders who want willing and able followers, not sycophants.

On the other hand, excessive compromise can give away too much. It can be seen as weakness, not as consensus-building. The wrong compromise can block progress not promote it. Like many people, I hate conflict. I shrink from confrontation. To avoid it, I am prone to give anything away. To escape the inevitable conflict I have refrained from removing people who were, by their incompetence, I knew, harming the organisation. I have been tempted to let a bullying neighbour have his way rather than confront him. That is bad compromise, compromise for the wrong reason. I was compromising the truth for the sake of a quiet life, a minor principle for a greater one. It should be the other way round, as I always knew.

Arriving at one management programme, the executives found on their desks an English translation of Sophocles' Greek tragedy, *Antigone*. This was their first homework of the course, to be discussed in class the next week. They thought at first that they had been mixed up with a liberal-arts course but they realised the point of it when they started to talk about it as a group. Antigone's brother had been defeated and killed by their uncle, Creon, in a battle for the control of Thebes. He had been left outside the city walls to be picked at by the vultures. Antigone's faith required her to see that her brother was properly buried, lest he be pursued for ever by the Furies. Creon forbade it, on pain of her death. Do you obey authority, or do you do what you think is right, come what may? Or do you find some compromise? It has been the dilemma of conquered peoples down the ages. It is a conflict not unknown in businesses, or even in families. Antigone stuck to her principles, and died for them. There *are* some principles worth dying for. The question is – was this one of them? It is because that question is forever topical that the play is still performed 2,500 years later.

Only when the compromise is in pursuit of a greater purpose, or a greater principle, is it right to compromise a principle. As well as conflict, I abhor war and violence. But there are bullies, and there are bully states, who respect only those who are stronger than them and who are prepared to show it. I will, therefore, condone limited wars and the controlled use of force in pursuit of peace and order, if all else fails. I will even fight myself, if the cause is just. To stick too tightly to my principles would be to give a licence to all bullies and all criminals. The greater principle of a just order legitimises the abandonment of a lesser principle.

Most compromise in life, however, is not about our principles but about our interests. No compromises on these can ultimately mean no allies, and no progress. The philosophy of Chinese contracts can then prove more fruitful if less glorious.

Contracts with the Future

Time also demands its compromises, as we try to balance the demands of the present and the future. Short-termism is an ugly word for a tough dilemma. Businesses are accused of it, governments are plagued by it, none of us can escape it in our own lives. We all live in the knowledge that those things which we most want, and which are best for us – health, affection, long life – all require us to give up immediate delights or to do things which we would rather not do. Personal short-termism damages our health. We know, in other words, that this sort of personal compromise is one way of dealing with the paradox that most of the things we enjoy are bad for us.

The dilemma is this – to what extent should you short-change or compromise the present in order to benefit the future? All investment involves taking something from today to improve tomorrow. It only makes sense to do that if you believe in, or want, what tomorrow may bring. It is always another compromise. To what extent are we prepared to curb our bad environmental habits to ensure a cleaner, safer world for our grandchildren, a world which we may not live long enough to enjoy? To what extent will we curb our own behaviour if others do not do likewise? Will the tragedy of the commons be played out on a global scale or will we compromise, adjust, our short-term behaviour for a greater common cause, to make life better for people we shall never meet? We will only do it, I believe, if we can look beyond the grave, if we can accept that there are some things that are more important than ourselves, and longer-lasting.

At a more personal level, young dual-career couples struggle with this issue of common cause and compromise, as they try to decide whether or when to start a family. The sacrifices in the present will be considerable: a loss of income, a change of life style, an altered relationship. The commitment of both of them to a new future is critical, if they are to make the compromises which will be necessary to the start of another family. It is, however, an impossible decision to make, if they want to preserve all that is in their present, while growing that future. They have to start by understanding that compromise is essential to most progress, but that voluntary compromise is only possible if there is a common cause, a cause greater than oneself, and a trust in the other. When compromise goes out of fashion among the young, so do babies.

In a business, to increase a dividend is also to reduce the sums available for new capital spending. If the shareholders are not interested in the future of the company because they can sell their shares tomorrow, they will want to see dividends, not retained profits, in the account. The managers, on the other hand, with their own futures linked to the future of the business, will want to invest as much as they can in that future. There can often be a conflict of priorities.

If there is no common cause, no agreement on the longer-term goal, the more pressing priority, or the most powerful party, will win out. If we think that we need shareholders more than managers, as we seem to, then the shareholders will win. Compromise will be enforced, not voluntary – a British contract rather than a Chinese one. Only if the shareholders are also locked in to the future of the business, as shareholders more often are in Japan and Germany, will they have common cause with the managers and be prepared to forego some present gains for future profits. As long, that is, as that common cause seems a worthwhile one. In the end, for the long-term to prevail over the short-term, we must want what the long-term promises. Where there is no vision, there you find short-termism, for there is, then, no reason for compromise today for an unknown tomorrow.

The concept of stock options, common in Britain and America, is an attempt to make common cause between senior managers and the shareholders. The thinking is that this will tie the managers' compensation more closely to the longer-term performance of the company. It does, but because they can only make use of those options after a period of years or if the share price goes higher than the price of the options. If all shareholders were treated this way they would look at the company a little differently. It would significantly alter the balance between the present and the future and so make compromise easier.

The Third Angle

If we want self-reinforcing relationships, those Chinese contracts, we need to find a common cause, one which justifies some personal sacrifice by both parties for a greater common good. Without that sense of common cause, everyone will fight their own corner, trust between individuals or groups will be rare and compromise will be something imposed on the weaker by the stronger, a sign of defeat not progress. In a democratic culture, if it is not to degenerate into a battle between interest groups, it is particularly crucial that we find that common cause.

Where there is a third corner, we can often agree to combine our individual corners against the common enemy. The calculation is that we have more to gain by combining than by standing alone. The value of the compromise seems self-evident. Common enemies, however, must, by their very nature, result in temporary combinations because the compromise was made on the assumption that the enemy would be defeated. When the defeat occurs, the alliance evaporates. If, on the other hand, the enemy continues to resist defeat, the compromises made for the alliance begin to be questioned when there is no pay-off. When businesses define their purpose as being bigger or better than one of their competitors in the hope that this will unite their workforce, they are playing short-term games. When the British government sets its sights on outperforming Germany or France, they are playing games too long to be real.

A common enemy is seldom a good basis for a lasting compromise or for long-term sacrifice, except in times of real, not economic, war; times which, we must hope, will seldom come. In many a situation there is, however, a third angle, a way of finding that balance between opposites which is a necessary Pathway through Paradox.

1793 was the bicentenary of one of the most influential graffiti of all time. On 30 June of that year, the Club des Cordeliers of Paris passed a resolution that all house-owners should be invited to paint on the façades of their houses, in big letters, these words, 'Unité, indivisibilité de la République, Liberté, Egalité, Fraternité ou la Mort'. It was the first recorded example of what was to be the motto of the French Revolution – Liberty, Equality and Fraternity. It was also the best-known example of trinitarian thinking, the 'third term' which reconciles opposites. Liberty, notoriously, cancels out equality, and vice versa. The two can only survive in any sort of harmony if there is fraternity. If we care for each other I will not press my demands for individual liberty so far that it intrudes on your equal right to be free, nor will your pressure for equality be pushed so far that it denies me my liberty.

Most of life is made up of opposites – male and female, work and leisure, life and death. In the British tradition it is always assumed that a conflict between opposites is the best recipe for fairness. The system of justice is based on this assumption, as is the British parliament. In the view of many it is neither fair nor always just.

Trinitarian, or third-angle, thinking is always looking for solutions which can reconcile or illuminate the opposites. A third party may yet be the answer to the see-saw of British politics, a see-saw that periodically gets jammed so that one party stays in power too long for fairness. An independent assessor in the judicial process may yet emerge as a solution to the battle of the courtrooms where victory for one side or the other is not always the same as justice. A common humanity is the concept which makes sense of the conflict between male and female. If we knew better what we meant by eternity we might not see life and death as such polar opposites. Learning might be a way of linking work and leisure so that they blend into each other. Love quenches rows and turns differences into stimulants; and sympathy, Adam Smith would remind us, makes the market moral.

Trinitarian thinking urges us always to be on the look-out for another approach, a third angle. Trinitarian thinking says that if money is so divisive, why not de-monetarise society? If more of the good or necessary things in life were free to all, like education, housing, health care, travel and essential foods, there would be more equality around. It would be less easy to distinguish between rich and poor, and there would be less reason to pile up one's riches. Trinitarian thinking suggests that if the struggle between 'owners' and 'workers' is endemic, we could get round it by thinking, instead, of 'members'.

I remember only too well the arguments with my two children over the telephone. We tried budgets, time-limits, locks and total prohibition as ways to curb their insatiable appetite for telephoning their friends as teenagers. It only resulted in evasion, lies and rows. Trinitarian thinking suggested another route, a third angle. I paid for them each to have their own telephone and paid the standing charge provided that they then paid the bills for all their calls. If the bills were not paid they got cut off. My phone would not then, or ever, be available. They saw this as freedom, but I noticed that, from then on, it was their friends, no doubt using their parents' phones, who telephoned them. My hugely reduced bills more than compensated for the extra standing charges, and, besides, I now had unrestricted access to my own telephone! It was a compromise which worked without any sacrifice of principle.

In one condominium, the problems of parking were the hardest problem to solve. It was seen as a basic right to be able to park your car, and that of your visitors, outside your apartment. On the other hand, no one wanted to have their view interrupted by other people's cars. Endless arrangements and architectural devices were explored but no one was satisfied. Trinitarian thinking suggested that the fairest solution would be to provide a communal parking site away from all the apartments, to declare the main area a car-free site and to landscape it appropriately. Some shared inconvenience became a small price to pay for enhanced beauty and peace. Another compromise which worked, because we found the third angle. Shrewd negotiators know all about such unblocking trinitarian devices.

This book will contain many examples of trinitarian thinking. It is often a path to acceptable compromise, a third angle which reconciles the opposites. However we get there, we need more Chinese contracts in our lives, in our businesses and in our societies. The Japanese are famed for the slowness of their decision-making but also for the strength and reliability of those decisions once made. The reason, as we know, is the length of time it takes to reach a consensus. You could say that the Chinese contract is a Japanese habit! The rest of us could learn it with advantage, but it will mean changing some of our cultural traditions: the idea, for instance, that winning necessarily means that someone loses, instead of the possibility that all can win a little less; that compromise is a sign of weakness not of strength; that a good lawyer is better than a good agreement; and that if you look after the present the future will take care of itself. These things are culturally determined, they are not engrained in human nature. We can change the way we think.

The starting premise of this book was that there is no perfect solution to anything, and that no one can predict the ultimate effect of any action. Only when you look back can you see the truth. Even science knows no perfect answers, nor believes that there can be any. Given this premise, one would have to be very arrogant, very stupid or very insensitive to claim to know the complete truth of anything in advance. Karl Popper once said, 'We all differ in what we know, but in our infinite ignorance we are all equal.' Even popes can be wrong, which won't stop them or many others from seeking to impose their wishes on the world.

We all have the right to think that we are right, because the good news from that starting premise is that there is room for each one of us to make a difference. Fluttering our wings, we too, like the butterfly in chaos theory, may cause an upset in the weather. Where no one knows the future with any certainty, we have the right to dream. But for anyone's wishes, or their dreams, to be accepted without enforcement, some form of Chinese contract will be necessary, one in which the interests of all parties, both now and in the future, are heard and heeded.

In turbulent times we look for certainty and sure authority. We want to be followers, not leaders, even in a small way. We want 'them' to solve our dilemmas, and give us back a quiet life. It is hard enough to look after oneself, without taking other people into account. I am arguing that these things cannot and will not be. The paradoxes are too complicated. We have to get involved if there is to be any point to our existence. In the last Part of the book I shall argue that there is a point, that we are not just incidental accidents in the evolutionary chain. In the meantime we have to put these principles into practice, in our work and in our lives. That is the concern of the next Part of the book.

Part Three: Practising the Preaching
Managing Paradox

The Federalist Idea

The symbol of the empty raincoat challenges us to find a place for our individual wishes and decisions in the greater scheme of things. Different cultures give a different prominence to the idea of the individual, but one can sense a growing sense of impotence, everywhere, in the face of institutions and government and transnational bodies. Democracy used to mean that the people had the power, but that now translates into 'the people have the vote', which is not the same thing. The vote is an expression of last resort, a useful reminder to our rulers of the source of their bread and butter, but hardly a way for an individual to influence what is going on around them. Moreover, in the institutions of everyday life, particularly those of business, the only people with the vote are those outside, the financiers or the governors. Those who work in them are effectively disenfranchised. Democracy has its limits. A vote is not enough to fill a raincoat.

If we want to reconcile our humanity with our economics, we have to find a way to give more influence to what is personal and local, so that we can each feel that we have a chance to make a difference, that we matter, along with those around us. We have no hope of charting a way through those paradoxes unless we feel able to take some personal responsibility for events. A formal democracy will not be enough. We have to find another way, by changing the structure of our institutions to give more power to the small and to the local. We have to do that, with all the untidiness which it entails, while still looking for efficiency, and the benefits of co-ordination and control. More is needed, therefore, than good intentions to 'empower' the individual to do what we want him or her to do. The structures and the systems have to change, to reflect a new balance of power. That means federalism.

Federalism is an old idea, but its time may have come again, because it has been designed to create a balance of power within an institution. It matches paradox with paradox. Federalism seeks to be both big in some things and small in others, to be centralised in some respects and decentralised in others. It aims to be local in its appeal and in many of its decisions, but national or even global in its scope. It endeavours to maximise independence, provided always that there is a necessary interdependence; to encourage difference, but within limits; it needs to maintain a strong centre, but one devoted to the service of the parts; it can and should be led from that centre but has to be managed by the parts. There is room in federalism for the small to influence the mighty, and for the individuals to flex their muscles.

We think of federalism as something which applies to countries – the United States of America, Germany, Switzerland, Australia, Canada. Her politicians might not admit it, but the United Kingdom is really a federation of its separate regions, as is Spain and, increasingly, even France, as its regions gain more autonomy. The concept, however, goes beyond countries. Every organisation of any size at all can be thought of in federal terms. Hospitals, schools, local government and most charities are, if we look at them with federal spectacles, made for federalism – local and separate activities bonded in one whole, served by a common centre. All businesses of any size have federal propensities, and a need to be all the things which federalism offers.

We might then wonder why such a good idea is not so obviously popular. Few businesses are consciously federal, nor does history have many, if any, examples of a monarch or a central power voluntarily moving to a federalist structure. The hard truth is that we are always reluctant to give up power unless we have to, and federalism is an exercise in the balancing of power. The federal idea is an example of the second curve, but one which too few institutions or societies develop until they are forced to. It is a very different and a very uncomfortable way of thinking about organisations. It is messy, untidy and always a little out of control. Its only justification is that there is no real alternative in a complicated world. No one person, or group, or executive, is so all-wise and so all-sensitive that they can balance the paradoxes on their own, or run the place from the centre, even if people were prepared to allow them to. We have to allow space for the small and the local.

Federalism relies on a set of Chinese contracts between its various parts and operates through doughnuts of varying size and shape, doughnuts which leave, of necessity and of right, considerable space for local decisions. The goals of the parts have to adjust to the requirements of the whole, and vice versa. No one in a federal organisation can have everything exactly as they would want it. It is, therefore, an excellent example of putting the preaching of this book into practice, with all the difficulties as well as the opportunities.

Federalism is not, let us be clear, the easiest of concepts to make work, or to understand. Yugoslavia is hardly an advertisement for the concept, nor is Canada. California is creaking under an excess of federalism from within and without. IBM proclaims its conversion to the idea, but may not be its most successful exponent in the years ahead. A federal Europe frightens many, not just in Britain. Nevertheless, we have to persevere with it, because it is the best way of returning some sense of meaning to our larger institutions, a way of connecting their purposes with their people.

Much of the confusion and difficulty arises from a misunderstanding of what federalism is. A confederation, for example, is not the same thing as a federation. A confederation is an alliance of interested parties who agree to do some things together. It is, therefore, a mechanism for mutual advantage. There is no reason for sacrifice or trade-offs or compromise, unless it is very obviously in one's own interest. A confederation is not an organisation that is going anywhere, because there is no mechanism nor will to decide what that anywhere might be. The Confederation of Independent States, which replaced the Soviet Union, will never be an effective body. The British Commonwealth, another confederation, is a thing of sentiment and language, not a real organisation. These are not the stuff of federalism.

Confederations adapt when they have to, usually too late. They do not lead, nor do they build. They are organisations of expediency not of common purpose. The British would like Europe to remain an economic confederation, a common market. Many in the rest of Europe want a more federal state, one with a greater common purpose, within which sacrifices and compromises become acceptable, one in which the rich are readier to help the poorer, one with common standards and common aspirations.

What is true of Europe is also true of organisations. Alliances, joint ventures and networks are the tools of confederations, arrangements of mutual convenience, inevitably fragile as the conveniences change. Organisations with a clear purpose will want to be federal not confederal. The distinction is important.

The key concepts in federalism are **twin citizenship** and **subsidiarity**. They are explained in the two chapters which follow. They are old ideas, reinvented for today's world.

6 Twin Citizenship

The Texan, when I visited him, was very obviously a Texan, and proud of it. On his front lawn, however, there was a tall white flagpole flying the Stars and Stripes. 'Why not?' he said, in response to my query. 'I'm an American, aren't I?' He was American *and* he was Texan. So, too, the lady from Munich, a Bavarian first, she said, but also a German.

Big but Small

Twin citizenship is a critical component of federalism: you belong to your own state and to the larger federal union of states. Each of us will probably have a number of twin-citizenship situations in our lives, if we think about them in that way. It is important that we do. Even in personal life one ends up with twin loyalties more often than not. When I married, I thought that I was joining two lives together, mine and that of the girl I loved. It was only when we started to arrange the wedding that I realised that I was also becoming the affiliated member of a new tribe, her large and extended family. I now had my own tribe and hers, twin citizenships and twin loyalties to be reconciled and balanced. Organisations, for their part, are going federal, just as countries have, because they want to give a measure of independence to local units or to specialised groups. At the same time, they want to retain the benefits of scale. Being big and small makes sense, but managing it makes problems.

Small units are faster, more focused, more flexible, more friendly and more fun – to borrow Rosabeth Moss Kanter's five 'f's. Small units can get closer to the customer and the citizen, to the patient or to the student. They can be less bureaucratic and more personal. Most of us fish prefer a smaller pond. In smaller groups there is more chance to be yourself, less likelihood of being an anonymous empty raincoat.

On the other hand, there are always economies of scale to be had in any organisation. They are most easily found in finance, distribution and purchasing; less often, these days, in manufacturing. Research, if it is any good, is increasingly expensive; it needs large cash flows to support it. Pharmaceutical companies have to be large, as do oil companies and aeroplane-makers. Large organisations offer better hopes of security to their people, more variety and more scope. They can afford to invest in the kind of fundamental training which takes years to pay off. 'Now that you've been with us for ten years,' my oil company told me, 'you may be worth your keep!' Many worry that if the BBC in Britain were to turn itself into a host of mini-independents, there would be no one with the interest in paying for the professional training which grooms all the directors and technicians of the future.

Federalism is an age-old device for keeping the proper balance between big and small. Big in some things, small in others. It is never easy, because it means allowing the small to be independent while still being part of the larger whole, to be different but part of the same. Federalism is, therefore, fraught with difficulty because it is trying to combine those two opposites, to manage the paradox. Twin citizenship makes it possible. If there is the sense of belonging to something bigger as well as to our own smaller unit, we can see the sense in accepting some restrictions on our local independence, if it helps the larger whole. Sovereignty is not ceded but shared. The larger unit is not 'them' but also 'us'.

To many Britons, Europe has always been a place one went to, not a place one belonged to. Until they instinctively realise that they are in it, geographically as well as politically, Europe and the European Community will remain 'them' not 'us'. There will be no sense of twin citizenship and therefore no gut feeling for federalism, which will remain a dirty word, a synonym for a loss of independence without any compensating new belonging.

The local citizenship is the easy part. We can all identify with our immediate neighbours, particularly when we are also working with them. Our futures then depend on them. We share a history, we know their faces, often too well; we know where we belong. In the words of an old Chinese proverb, those who do not know the village they have come from will never find the village they are looking for. Richard Rogers, the architect who designed the Pompidou Centre in Paris and the Lloyd's building in London, has commented that Europe is becoming increasingly defined by its cities. Barcelona's Olympics were about Barcelona not Spain. Dresden is intent on being, once again, a centre of culture, art and education, a place for Europe not just Germany. We can relate to our cities more easily than we can to the idea of Europe, or America.

It is that larger citizenship which is harder to establish, yet it needs equal prominence if there is to be a proper balance. In a business, it may make good logical sense to combine functions, to group some regions together, to manage cash or purchasing centrally, but these actions all steal power and decisions from the independent units. Those units will resent this, not understanding the paradox that in order to get the most value out of their independence, it often pays to sacrifice some of that independence to a central function. That kind of compromise is only done willingly if there is some confidence in the central function, some sense of belonging to a larger whole of which it, too, is part. We need that second citizenship.

The £5 Auction

To prove this obvious point I used to play a simple game with executives on training exercises. It was a variety of what logicians call the Prisoners' Dilemma, except that in this game I offered to auction three £5 notes between two participants. I would place two volunteers in chairs facing away from each other so that they could not see the other's face. I asked them to bid in turn for the first note. Invariably the first person bid £1, the second upped it, and so it went on, by alternative bids, sometimes going up by £1, sometimes jumping by £2, sometimes by parts of a pound, until one of them reached £5 at which point the other normally, but not always, stopped bidding. In one case, someone actually bid £23 for the first £5 note! The auction for the second note gave first bid to the other side but the outcome was the same – £5 or more bid for my £5 note. So it was for the third note, although there was often overbidding for this note, too, so that one side could claim a sort of Pyrrhic victory – more notes won, and to hell with the cost!

The rest of the group would be watching, amazed by the apparent idiocy of the bidding. There would be a rush of volunteers for the next round, eager to try their theory of pre-emptive bidding or whatever. The result would be the same, as long as I was careful to pick them from different sides of the room. Finally, I would choose a couple who had been sitting and whispering together and who volunteered in unison. When they started the bidding, the first person would bid 10 pence and the second would say 'no bid'. The note had been sold for 10p. The same happened the other way round with the second note. The third note was more tense. Usually their agreement held. The first person bid the now-standard 10p and the other passed. At the end, they took the three £5 notes, paid me 30p and shared the proceeds. Occasionally, however, competition flared up again for this final note and the second bidder would come in with a pre-emptive £5 bid. They would then have to make do with a £4.90p profit and live with a sense of betrayal.

What was going on? I would ask the group. Logical, sensible, mature individuals were competing to the point of lunacy because I had kept them apart. By not allowing them to communicate, I had also prevented them from establishing an alliance, an agreed objective and a means of proceeding. Only when I picked people who had had a chance to talk together were they able to achieve a common goal which benefited them both, although even that broke down on occasion. A common cause, the willingness to deny oneself in the interests of that common cause, and the trust that the other party will do the same – these are the essentials of sensible organisation behaviour. Much of the time this sensible behaviour does not happen because people do not talk, do not trust and have no common cause. To put it more crisply, there is no sense of a second citizenship, and therefore no possibility of sensible compromise, of a proper balance between the whole and the parts.

The depressing thing was that the experiment never failed. It always worked the way I knew it would. We instinctively work for our own immediate advantage unless there is an obvious common cause with people whom we can trust so that an initial sacrifice turns out in the end to be to our mutual advantage. We can, today, see the £5-auction game being played for real around the world.

The Queen's Great Matter

The second citizenship is critical. Interestingly, politicians and managers both use the same sort of devices to reinforce that larger loyalty. They make sure that there is a federal flag, or company logo, which is displayed wherever and whenever possible. They have their national anthems, or vision and value statements in organisations, things more symbolic than real but important none the less because they give expression to the ideal which holds the whole together, the sense of common cause to which we are, they hope, committed.

Modern organisations spend a lot of time working on that common cause, establishing what it is, communicating it, reinforcing it. It can look like waffle, and it sometimes is. Properly done, it is not waffle but the glue of the enterprise. More respectfully, Richard Pascale and Anthony Athos call it the 'spiritual fabric' of the corporation. They are describing companies in present-day Japan. The tradition is older still. In Elizabethan England they called it the Queen's Great Matter, the common cause that bonded her merchant venturers and built an empire. In America, every president goes out of his way, in his inaugural address, to emphasise that he will now govern for all his fellow citizens, not just the ones who voted for him. He points out that they can only help themselves if they also help each other. He reminds his audience that there is this place, this tradition called America which they need to rediscover. 'We must prove that we deserve our heritage.' This is not empty rhetoric but the necessary forging of a common purpose. Sadly, the rhetoric too easily gets lost in the day-to-day reality. Living up to the rhetoric of the larger citizenship is one of the toughest parts of leadership.

Presidents, leaders, to be effective have to represent the whole to the parts and to the world outside. They may live in the centre but they must not be the centre. To reinforce the common cause they must be a constant teacher, ever travelling, ever talking, ever listening, the chief missionary of the common cause. This role sits ill with that of chief executive, which is why many organisations are now separating out the two roles. The missionary task is a role which cannot easily be fulfilled by committee or by memorandum because logic makes few hearts beat faster and no one has ever followed a committee into battle. The life of the federal president in a large organisation tends to be one long teach-in. Successful presidents and prime ministers know that their main task is to carry the people with them. Roosevelt with his fireside radio talks, Clinton with his town meetings, Churchill and his wartime broadcasts, were all, in effect, running popular teach-ins.

Twin citizenship, however, needs more than flags, national anthems and articulate, visible presidents. It is lubricated by cross-fertilisation, by moving people between the parts, and between the parts and the centre. In that way more people are exposed to more of the bigger reality; they not only grow themselves, but their vision and their understanding of the total organisation grows with them. Shell, one of the oldest of the corporate federations, knows that its corps of 5,000 expatriates is the bond which holds it all together, far more than any shareholdings or formal authorities. On a different scale, the Club Méditerranée, another federal organisation with its independent holiday centres, insists that the site managers change locations every two or three years, creating an international family of shared habits and values.

The Law and the Currency

A common legal framework and a common currency are the other essentials. They both remind everyone, constantly, that they are part of something bigger. That is their symbolic role; they also have a practical one, to allow the parts to work together.

A common legal framework translates, in organisations, into a basic set of guidelines – 'how we do things around here'. They call it the 'bible' in some places, but they would be wise not to make it as bulky or as subject to devious interpretation as the Bible proper. Too much law only breeds lawyers, who come expensive. A common currency translates into a common information system, so that inputs and outputs can be measured and compared across the parts. That sounds like common sense were it not so uncommon. Too often the sales section talks of sales but knows not what contribution they make to added value, while purchasing talks costs but knows not what difference they make to sales. In the past, in Britain, when doctors prescribed drugs or referred their patients to specialists, they neither knew nor thought it right to know the cost. To know would, they felt, have influenced their medical judgement. They then complained that the money to cure their sick was never adequate. In the National Health Service at that time there was no common currency and, therefore, no sense of involvement in, or responsibility for, the financial aspects of the organisation. Paradoxically, by breaking down the health-care system into more self-accountable units, the sense of common cause has become stronger because the parts now have to talk to each other and have the means to do it. As they say, an organisation that talks together stays together.

We must, however, take care that the laws and the currency are not so pervasive that they swamp the local citizenship. The centre may yearn for uniformity, for an identikit organisation in which every bit is the same as every other bit, but the customer or client wants it to be his or her preference at the point of delivery and the local unit wants to be able to be appropriately different. Uniformity, like equality, is the enemy of liberty, but too much liberty can destroy efficiency. There has to be a balance.

Our new local superstore does not stock risotto rice. Risotto rice might be thought exotic in an out-of-the-way corner of England but I delight in the infinite variety that is possible with this simplest of ingredients. We asked for it to be stocked. I spoke with the store manager. He was regretful, but the shelf lay-out and the stock-list were laid down from the centre and he had no discretion. How, I wondered, could the centre be so all-wise as to know the rich randomness of tastes in our neck of the woods, or did they not trust him? What, if any, discretion was allowed to him in other areas or was he but a walking automaton, there to give standard non-responses to people like me? There was little sense of local citizenship there. An identikit, facsimile organisation imposes uniformity on its world, thinks that the centre knows best, that discretion is dangerous and that local differences are unnecessary. All those assumptions, I suggest, are dangerous because they deny the paradoxes, they bury them instead of balancing them.

Money Talks

We would be wise, too, to remember another lesson from that auction game which I played with my students. Money talks. Money may not be the most important thing in the world but it provides the counters. Twin citizenship is much easier to believe in if there is some financial involvement in both citizenships. Without the tokens the game may not seem to be real. Our pay is beginning to reflect this. Profit-sharing schemes are not as rare as they once were. They may go farther still. The time will come when an individual will find that the annual pay packet comes in four lumps. The largest lump will be the pay for the job, reflecting one's standing in the organisation, one's level of experience, expertise and previous record as well as the level or grade of job. That is nothing new. What would be new, except in Japan, is that this lump might be only 50 per cent of the total take-home pay in a good year. The other lumps would be a share in the overall surplus of the group or corporation, a share in the value added by one's work unit, the first citizenship, and, finally, a personal bonus reflecting one's individual contribution. The two shares might normally be expected to amount to 20 per cent of the total packet leaving 10 per cent for the individual contribution.

At first sight these numbers seem huge. Remember, however, that they start with a base salary or wage which is set at 50 per cent of normal take-home pay. That sum will not vary but the other numbers will, in line with actual performance. When times are good, the money will be good and there will be that much more to distribute because the basic costs are only half of what they could be. When times are hard, however, or performance slips, then total pay declines, but no one need be dismissed to reduce the labour costs; they reduce automatically. In that way the obligations of citizenship are met – you keep your job – and the rewards are shared when there are rewards to share. It is this kind of payment system which has helped Japanese companies to maintain their system of lifetime employment for their key workers.

The percentages need to be big to be interesting. A total bonus element of 5–8 per cent, the kind that some organisations play around with, does little more than pay for the Christmas or the New Year break. It is a gesture not a bond. The percentages must also be seen to be fair and objective. They must be based on real numbers, not on percentages or judgements. The exception may be the percentage allowed for individual contribution. This can seldom be totally objective unless one is a salesperson on commission. Opinions have to count. The sense of local citizenship will be reinforced if the opinions are those of the group itself, which is given a total sum of money to distribute amongst its members. A good, open group will not flinch from the task of allocating the money, accepting that they will have to give their reasons for giving anyone less or more than the mean. On the other hand, one's work group is not the only arbiter of one's contribution; superiors and colleagues outside the group have relevant opinions, while many a group will duck the issue and will share the money out equally. Balance is best. Opinions from both inside and outside the group also increase the opportunity of feedback. Critically, however, this individual element in the bonus should be less than the two group elements if the sense of twin citizenship is to be reinforced.

Obviously, such a dramatic rearrangement of the reward system in an organisation could only be instituted in one step on a greenfield site. Established organisations will have to get there more gradually, taking full advantage of good years to move forward faster, never forgetting the ultimate goal and being careful to explain, at all times, the why and what of it all. It has to be seen as a way of sharing in the rewards of citizenship as well as in its risks.

The Disappearing Middle

Twin citizenship implies that we are citizens of only two states. It should in theory be many more. I am a citizen of my town, then of my region, then of my country. Above and beyond that comes the trading bloc or the larger federal state and beyond that, why not, the world. Theory, however, does not always sit easily with psychological reality. Most of us seem to be capable of only two levels of loyalty in any one area of our lives. The ends of the chain, therefore, often get dropped and the centre levels get squeezed out. 'I am a Scotsman first,' my friend said, 'and then a European. I don't feel British at all.' Only those countries which are, like my Ireland, small enough to be tribes, do not get squeezed between the tribe and the federal union. Businesses which try to put another layer of loyalty, often a geographical one, between the operating company and the corporate centre, can end up confusing and weakening the sense of citizenship.

Governments, in their turn, have to decide which layer

to omit in dispensing health, education or welfare to their citizens. If they insist on retaining the central control while delegating delivery to local units, the intermediary levels will only get in the way and will atrophy. Either the national government has to shed its power to the intermediary level, retaining only the roles of service- and advice-provider, with money allocated according to formula, or it turns the intermediary levels into optional resource centres for the units to draw on if they wish. Federalism has then become a mechanism for centralisation.

A more interesting example of the disappearing middle in government is suggested by David Osborne and Ted Gaebler in their book *Re-inventing Government*. They want to see more of the ownership and control of public-service institutions passed out of the hands of bureaucrats and government professionals into communities and individuals. Citizen groups, neighbourhoods, volunteer organisations would be authorised, and, where necessary, be centrally funded, to carry out many of the local activities of government. If that were to happen, whole layers of administration would be unnecessary.

The new executive agencies in Britain are a step along this route. These are autonomous entities charged with the delivery of government services, ranging from the Benefit Agency to the Central Office of Information. When they are truly autonomous, the federalism of government services will be well established. At present, the British Treasury is still reluctant to let go all the strings. The pay and the grading and the numbers of staff, for instance, are centrally controlled. There cannot be a true feeling of local citizenship when you cannot determine your own staffing levels. Nor is the monitoring of technical details a substitute for the second and bigger loyalty which will be essential if the old traditions of the Civil Service are not to get lost in the new proliferation of independent bodies. Federalism is not easy.

There can, however, be too many definitions of local citizenship. The State of California is fast becoming bogged down in too many layers of government. It is hard to know where real responsibility lies, what with school and hospital boards, local communities, the state and the federal levels, and the continuing experiment with direct democracy whereby the voters can vote to make specific propositions into law. Too many layers of citizenship end up as a bureaucratic nightmare, be it in a corporation or a country. More of the middle needs to disappear.

National parliaments in Europe's larger countries, which are themselves federations of tribal regions, know that they are likely to be squeezed out if and when Europe becomes a fuller federation. Understandably, they do not relish the thought. It is not nice to be a disappearing middle, even if a greater loyalty requires it. It is not only national parliaments which face this dilemma of the disappearing middle; layers in organisations have been collapsing for a decade at least, not least because those organisations are reorganising federally, even if they do not always call it that or recognise it as such. In the federal structure, hierarchies are limited and local; you relate to people in the wider organisation because their role is relevant to your needs, not because their status in the organisation requires it. Forget the hierarchy, use the network.

I listened to the chairman of a large French supermarket and hotel chain explaining his federal, devolved organisation to a sceptical Spanish audience who were still hooked on hierarchy. 'Please explain to us,' one of them eventually asked in some frustration, 'to whom does the manager of the store in Lyons report?' The chairman clearly did not understand the question: 'Well,' he said, 'if it's a question of distribution he will go to the expert who is, I think, in Marseilles, but if it is a purchasing problem the right person is in Paris.' 'Yes, but who is his immediate boss?' 'There isn't any one person whom he would call "boss".' You could see the mystification on the faces of the Spaniards who live in a partially federal country but do not, yet, run their organisations federally. Twin citizenship needs no middles.

The Loss of Loyalty

Twin citizenship is key to one set of paradoxes, in our societies and in our organisations, be they businesses, hospitals, government agencies, charities or whatever. Deny the local smaller loyalty and we kill all liberty, incentive and initiative and rely on the centre to be right, as IBM did, to its great cost, in the early 1990s. Deny the bigger loyalty, and inefficiencies, duplications and misunderstandings will proliferate. We need both loyalties.

In 1993 the Social Affairs Unit of the British Home Office produced a book of essays entitled *The Loss of Virtue* which argued that words like 'Duty', 'Loyalty' and 'Obligation' had disappeared from common usage. They blamed it on the growth of an amoral culture. Some attributed it to the failure of religion to cultivate a sense of right and wrong. Tony Blair, the Labour Party spokesman, then, on Home Affairs, pointed out that the words had no meaning unless one felt that one belonged to something, something which one could draw from as well as give to, something bigger than one's gang or club. He was, in my words, saying that, without a sense of a second and a bigger citizenship, selfishness is inevitable. If we cannot create that feeling of a bigger citizenship in our people, there will be no balance in society and our language will, indeed, begin to change.

Federalism, properly understood, can restore that sense of a local belonging and a broader, bigger citizenship, both in our organisation and in society.

7 Subsidiarity

Subsidiarity is an ugly word. But once you have learnt how to spell it and get your tongue around it, you will be unlikely to forget it. Subsidiarity is the idea at the centre of federalism; it is the key element in learning; change, if it is to be effective, depends upon it; the work of teams requires it, as does any attempt to make individuals take more responsibility for themselves. It is, however, a confusing word, because it has nothing to do with subsidiaries.

Reverse Delegation

Jacques Delors once offered a prize for a good definition of this ugly word. He need not have bothered, as various people were quick to remind him. Politically, the tenth amendment to the US Constitution, laying down the principle of States' rights, does it, without using the actual word. Much earlier, the Roman Catholic Church, borrowing the idea from political theory, coined the word and turned it into a moral principle. It was last restated in a papal encyclical, Quadragesimo Anno, in 1941: 'It is an injustice, a grave evil and a disturbance of right order for a large and higher organisation to arrogate to itself functions which can be performed efficiently by smaller and lower bodies . . . ' Strong words. I translate them more simply – stealing people's responsibilities is wrong. You could also define subsidiarity as 'reverse delegation' – the delegation by the parts to the centre.

Not so long ago my young daughter started her own

business with a partner. They had a good product, but they had never run a business before. As I watched them making what I was sure were dangerous and foolhardy decisions, the temptation to intervene and give them the benefit of my experience was overwhelming. I loved my daughter and I badly wanted this venture to succeed. I wanted to help. I was bluntly told to mind my own business not theirs. I realised, belatedly, that I was stealing her decisions, taking away their choices and their chance either to claim success as theirs or to learn from their failure. I apologised. Next time I would wait for them to ask – reverse delegation. I understood, then, why subsidiarity was a moral principle.

Federal organisations take subsidiarity seriously. They have to because they work on the principle of reverse delegation. The individual parts, or states, cede some of their powers to the centre because they believe that the centre can do some things better on a collective basis than they can on their own. They therefore retain as much independence as they think that they can handle. These 'reserve powers' of the centre are negotiated jointly and are then recorded in a formal constitution. All federal organisations have written constitutions. It may be that Britain's aversion to a written constitution has something to do with her intuitive distrust of federalism and its formality. There should be nothing vague or woolly about federalism or the place gets cluttered up with overlapping responsibilities and misunderstandings.

As more and more organisations collect alliances around their cores, they are forced to negotiate what should be done by whom and the pressure will be to allow as much discretion to the parts as is sensible and possible. What you do not own you cannot dictate to; negotiation is inevitable, so is subsidiarity – leaving power as close to the action as possible.

The New Centre

Homa Bahrami, describing the new hi-tech organisations of Silicon Valley, calls them multi-polar, saying that they:

> are more akin to a federation or constellation of business units that are typically interdependent, relying on one another for critical expertise and know-how. They have a peer relationship with the centre. The centre's role is to orchestrate the broad strategic vision, develop the shared administrative and organisational infrastructure, and create the cultural glue which can create synergies.

One company employs 100 professionals in roles which are classified as 'corporate'. These include finance and administration, infrastructure support (which includes purchasing), legal services, human resources and corporate communications. All these roles, we should note, are service roles rather than decision-imposing roles.

The 'horizontal organisation' is also in fashion. As described by McKinsey consultants Ostroff and Smith, these organisations have ten key principles, including: 'organise work around processes not functions and select key performance objectives, flatten hierarchy by minimizing non-added-value activities, make teams not individuals the principal building blocks of organisations'. What they are saying is that the trick is to find the optimum level of subsidiarity and then collapse as much into that as possible so that the group or team or individual have the means at their direct disposal to do what they are responsible for. In their view it is the team which is closest to the action that is the appropriate level of subsidiarity. That done, it is the job of the centre to set standards but not necessarily to specify how they should be delivered. The unit is then judged, after the event, by its performance against those objective standards. Some call all this 'process re-engineering', but that is only to give a modern name to an ancient principle, a principle which needs to be rediscovered, if we are going to have any chance of coping with the turbulence of the times. No longer do people believe that the centre or the top necessarily knows best; no longer can the leaders do all the thinking for the rest; no longer do people want them to.

Following this principle, organisations, everywhere, have been collapsing and dispersing their centres. The 100 professionals of Silicon Valley seem to be about standard. ABB, the Swedish–Swiss engineering giant, oversees 225,000 people with about that number in an undistinguished office building in Zurich. British Petroleum, in London, has twice that number but would like to make it less. Richard Branson's Virgin empire makes do with five! One way that they do it is by dispersing the centre. There is no need to have all the people with responsibilities across the organisation sitting in the same central place. Those who are responsible for co-ordinating a particular product range may sensibly be located in the place which does most of the work on that product. The research co-ordination can go to the biggest laboratory, a geographical watching brief to a country or state in that area. It spreads power around and down. That gives those who are nearer to the action a sense of involvement, of ownership, in the policies of the whole. It is subsidiarity in practice, as it is when they locate the European Bank for Reconstruction and Development in London or the European University in Florence, or the European Parliament in Strasbourg. To put everything in Brussels would be to take too much power into the centre. It would be stealing responsibility.

Small the centre should be, and partially dispersed, but it must be strong and well informed. The centre, after all, carries the ultimate responsibility for the whole. Its reserve powers typically include 'new money', i.e. the choice of strategic investments; 'new people', i.e. the right to make the key personnel decisions in the group; the design and management of the information system, which is the artery of the organisation; and, most controversially, the 'right of invasion' when things go wrong. Only those in the centre can have a view of the whole. They cannot run it, and should be too few in number to be tempted, but they can nudge, influence and, if they have to, interfere. The centre's principle task is to be the trustee of the future, but it needs to be sure that the present does not run out before the future arrives.

Federalism, insists Mike Bett of Britain's BT, cannot work without a strong centre. In the past, this meant that the strong centre was also a big centre. In order to co-ordinate plans and monitor activities, a lot of people had to be around. Power was then concentrated in one place. Federalism existed in name only. The information revolution which has overtaken us means that the centre can now be well informed but small, it can be strong but dispersed. Power can be more balanced. The nerve centre of the organisation can be in the chief executive's laptop computer – and in several others simultaneously. The 'Virtual Organisation' – the image of the organisation on your screen – is almost here, in our briefcases. The information age has made federalism possible.

This new, dispersed, centre has still, however, got to talk to itself as well as contemplate its screens. Video-conferences, voice mail and other technological devices help, but there is no real substitute for looking someone in the eye while you talk or they talk. Dispersed centres mean a lot of travel and red-eyes. The physical centres of these new dispersed organisations increasingly begin to resemble clubhouses, places where people meet, eat and greet but do not do their daily work. Like a club, there is a resident staff, those corporate services listed above, for instance, but the key players in the organisation live and work elsewhere and use the club for their necessary meetings. It is not even essential that the chairman or chief executive works out of the central club. For a large part of their time the officers will be, anyway, out and about, with the troops, where the different decision centres are, teaching, coaching, looking, listening. When they do go to the 'club', they can even have their own up-market 'puppy' or 'cart', a mobile desk with all its electronic paraphernalia, which is wheeled out and plugged in whenever the owner checks in.

Italian Style

One begins to wonder, then, what will happen to the cathedrals of corporate power, the tower blocks which shape our skylines. Centres of 100 professionals and corporate clubhouses do not need rows of little boxes piled high into the sky. It has often seemed strangely appropriate that the executive suite should be so high that it is, on occasion, above the clouds, but now that it is recognised that those in the centre are not all-seeing we may find them coming, physically and metaphorically, closer to the ground. Will their old suites become apartments for rich geriatrics or will the whole edifice be pulled down? A changing skyline will be the outward sign of real subsidiarity.

The skylines of most Italian towns have not changed for centuries. It would, of course, be a cultural crime to tamper with the roofscapes of Siena or Florence, Rome or Bologna; but I suspect that the organisation of Italian society has something to do with it. In Italy, much to the frustration of its central government, real power still resides in the family and the local community. Subsidiarity has always bypassed the formal institutions of government. After I went to live, for part of the year, in Tuscany, I soon realised that there was no way that I could conform to the myriad Italian laws regulating the buildings you could build, the cars you could buy, the permits you should have, the taxes you ought to pay, the people you could employ and the people you might allow to rent your home. Not only do the regulations change rather frequently, but the bureaucracy cannot cope with anyone, such as a law-abiding Anglo-Saxon, who wants to do it all by the book.

Nobody, I began to understand, expects you to abide by the letter of every small regulation, but should you fall out with the local community they have an array of laws to throw at you. The community where we live is a network of families. Everybody knows everybody, and knows what everybody is up to. You disregard the locals at your peril, with the law as a weapon only of last resort. Outsiders are welcomed but will never be insiders. Government can pontificate, legislate and regulate but much of it has no effect. It is a very effective but informal system of local control.

Many would argue that, in Italy, subsidiarity has gone too far. The country is broke, and may, conceivably, be split into two or even three, while the locals seem to thrive. The Mafia, the biggest of the families, still rules in parts. Government is impotent, and has proved to be corrupt. The power has to be rebalanced if the country is to be a viable entity. This will only happen, however, if there is a general recognition that some powers have to be ceded to the centre for the good of all, because federalism depends upon reverse delegation. This recognition is slowly dawning; the old politicians who let subsidiarity run riot, often to their own advantage, are on the way out. The balance will, I hope, soon be restored.

Italy is a land of families, of small units linked by networks, in business as well as life. What the Italians do instinctively, we must do deliberately.

Subsidiarity means small units, small units with real responsibilities. Richard Branson likes units of 50 or 60, Anthony Jay, in *Corporation Man*, favoured 400 or 500 and provided evidence from schools, Paris suburbs and Australia. Bill Gates of Microsoft likes 200 as a maximum. Tom

Peters has documented many cases of organisations like Union Pacific Railroad breaking themselves up into smaller units, in that case of 600 people each, but comes down in favour of 150 as the natural size. He cites the findings in the *New Scientist* magazine that 'in most modern armies the smallest independent unit normally numbers 130–150 men', that 'there is a critical threshold in the region of 150–200, with larger companies suffering a disproportionate amount of absenteeism and sickness', that 'once an academic discipline becomes larger than [200 researchers] it breaks into two sub-disciplines', that 'neolithic villages from the Middle East around 6000 BC typically seemed to have contained 120–150 people' and 'the Hutterites, contemporary North American fundamentalists, regard 150 as the maximum size for their communities'.

Forget the precise size. The point is that we need the unit to be big enough to be competent to do what it has to do and small enough so that we can know everyone in it and they can know us. The Bishop of Occam would have understood. According to the principle of Occam's Razor, the unit should be as small as it can be and as large as it has to be, a paradox in balance.

Signatures and Rowing Eights

Subsidiarity, however, depends on a mutual confidence. Those in the centre have to have confidence in the unit, while the unit has to have confidence in the centre and the members of the unit have to have confidence in each other. When the mutual confidence exists, there is no need for the books of procedures, the manuals, inspectors, performance numbers and counter-signatures which clutter up large organisations. These are all the signs of distrust, the atmosphere of fear which makes so many organisations seem like prisons for the human soul. They should not, need not, be like that. Our work can be our pride. Put it this way: we want to be able to sign our own work. A lot of people already figuratively do sign their work. Every member of the team that makes a television programme puts their name on the end of it. As you watch the credits roll you wonder why anyone needs to know all those names. You don't need to know, but they need to tell you; they want the acknowledgement and the credit.

A friend was appointed manager of a small art-printing works. Shortly after his arrival he called the whole workforce together and told them that he was ashamed of the quality of much of the work that had been going out of the place. 'In future,' he told them, 'I want everyone who has worked on an order to sign their names to a slip that will go out with the order saying, "We are responsible for this work. We hope that you are pleased with it." I expected a revolt,' he said, 'but instead they cheered.' 'We, too, have been ashamed of much of the work. But we thought that that was what you wanted – the lowest acceptable quality at the lowest cost. We are happy to sign our names provided you supply us with the machines to allow us to do work to our standards.' Subsidiarity depends on a mutual confidence, but putting your name to it is the best guarantee of quality that I know. It is the reason why professionals always sign their work. The signature acknowledges their responsibility. We know who to blame if things go wrong – and who to thank if they go right.

Such mutual confidence, however, takes time to build up. It has to be earned by all concerned. I once described a typical British work team as being like a rowing eight – eight people going backwards as fast as they can, without talking to each other, commanded by the one person who can't row. I thought that it was witty. I was quietly rebuked, afterwards, by a member of the audience who happened to be an oarsman. 'You couldn't be more wrong,' he said, 'to make fun of it. We couldn't go backwards without talking, or be content to be commanded by a non-rower, if we did not know each other very well and have complete confidence in each other's ability to do the job we are supposed to do. That's why we practise together so much, eat together and even live together for long periods.'

I remembered, then, that Japanese groups are renowned for the time they spend together off the job, and I notice how actors not only rehearse together but socialise together. You have to know each other well, it seems, both on and off the job, to know whether you can have confidence, or more, trust, in someone. My son went through a typical British education, one designed to bring him out as an individual and to emphasise his personal qualities and skills. He stood out in a crowd. Then he went to drama school, where they select a group of 27 young people to work together, learning to perform plays, for three years. He quickly realised that he stood or fell according to the quality of the group as a whole. It is no use being a star in a mediocre team. He became a devoted groupie, teaching others what he knew and they didn't, and learning, in turn, new skills from them. Competition was out, co-operation was all. He had no time, he said, to see his other friends during the terms. The group came first. 'We all depend upon each other.' I accused him of going Japanese. 'That would be a compliment,' he replied, 'because they understand what is needed in a group.'

Tough Trust

Subsidiarity in a group sounds warm and reassuring. It is, in practice, tough, and has important consequences for those in charge. For one thing, the group has to be small enough and be together long enough for the mutual confidence to grow. Confidence and trust cannot be ordered up from the store. A person must remain in a post long enough for others to judge the consequences of their actions and decisions. One-year assignments will seldom cover the feedback loop. More important even than that is the necessity to be ruthless if the confidence turns out to be unjustified. If you cannot have confidence in a member of the team, that member must go. If the whole team does not merit confidence, the team must go. Without mutual confidence the principle of subsidiarity cannot work. Checks and checkers have to be installed. Suspicion and evasion become rife, morale declines, the work deteriorates and any remaining confidence evaporates. Mistakes can and should be tolerated, provided one learns from them, but too many mistakes erode confidence, particularly if they are what one company, W. L. Gore, calls below-the-waterline mistakes, mistakes which imperil the organisation. Those are not easily forgiven. It is better, then, to be tough than sorry.

Tom Peters tells a nice story of Mike Walsh taking on the job of turning around Teneco. Four months into the job he heard that local managers at the Louisiana site had called the employees together for a 'safety meeting'. When the workers arrived they were told to lie on the ground and were searched for drugs. This, felt Walsh, was not going to help the kind of organisation he was trying to create. He flew to the site, apologised for the search, and used the occasion for another general meeting. During the meeting some employees started to complain about a safety problem in some of the bunk cars where the employees lived when on the site. Local managers started to explain away the problem by detailing how much the company spent on bunk-car maintenance. Walsh interrupted. 'Why not just visit the cars?' 'But it's raining outside,' some of the managers said. 'It's OK,' Walsh told them, 'managers won't melt.' He visited the bunk cars, decided that they were indeed unsafe, and saw to it that they were fixed.

That action may not have seemed very important in the great scheme of things at Teneco, but stories spread, and that kind of action helped greatly to establish the kind of two-way confidence which is essential to subsidiarity. Confidence depends on knowing who the other person is, what they stand for, how far they will go – on basic human qualities like authenticity, integrity, character. These are a far cry from the spreadsheets and committees which permeate organisational life. Tom Peters devotes a whole chapter in his book to 'The Missing X-Factor: Trust', but has no easy solutions to offer. 'Read more novels and fewer business books,' he says. 'Relationships really are all there is.'

Subsidiarity sounds like another ugly word – empowerment. There is a significant difference. Empowerment implies that someone on high is giving away power. Subsidiarity, on the other hand, implies that the power properly belongs, in the first place, lower down or farther out. You take it away as a last resort. Those in the centre are the servants of the parts. The task of the centre, and of any leader, is to help the individual or the group to live up to their responsibilities, to enable them to deserve their subsidiarity. In this way it is possible to handle one of the paradoxes of individualism, that we want to belong but we don't want to be bossed around, or to be 'empowered' if the hidden message is 'I empower you to do this, but I can disempower you if I don't like the way you do it'. Subsidiarity, therefore, is a tough deal. One has to understand one's responsibilities and then deliver. It means, too, that we have to be able to face up to disagreements. If we are going to take responsibility we need to be clear about what the criteria for our success are to be, what is acceptable and what is not. Only if there is a mutual confidence can disagreement, argument and conflict be handled positively. Organisations based on subsidiarity are full of ambiguity and argument and conflict, but if it is argument among trusted friends, united by a common purpose, then it is useful argument. Truth, said the Scottish philosopher David Hume, springs from arguments among friends.

It is, as a result, extremely demanding to run an organisation on the basis of reverse delegation and confidence. It also feels quite lonely at the centre. As one director of ABB commented, 'All we can do is to watch the herd and observe, with some relief, that in general it is heading in a westerly direction!' Why, then, are so many organisations trying to make reverse delegation work? It is partly in response to the paradox of individualism, the recognition that the well-educated knowledge-worker increasingly wants both freedom and structure. To attract and keep the best of these knowledge-workers, to be a so-called 'preferred organisation', subsidiarity has to be guaranteed.

Most of us are little different from the knowledge-worker. We want to own our work, but we like to work within a structure. We need to know what is expected of us but then to have the discretion to do it our way. Subsidiarity is also, and more urgently, a response to the need in our institutions to be flexible but coherent, to be all things to all people but still recognisably the same to all. Deep down, however, subsidiarity is a moral imperative. Power belongs to the people. It is the manager's, or teacher's, or parent's challenge to help them to exercise it responsibly.

Subsidiarity, with its emphasis on our individual rights and duties, is the basis of any concept of citizenship and critical to any concept of society. If we want our personal freedoms, and if we want them underwritten with guarantees of health care and welfare, we must accept our responsibilities to our fellows and earn the confidence which will allow the freedoms. That is the kind of thing one learns from parents as much as from teachers, but, then, the messages implicit in subsidiarity are a good guide to parenthood. Give a child as much responsibility as she or he can handle and then help them to live up to it. Subsidiarity is an old word, packed with meaning. It may sound out-of-date but it carries a modern punch. We would be foolish to discard it.

The Meaning of Business

A book which starts with the symbol of an empty raincoat and the challenge to find our human selves again, amid all the pressures for progress and economic success, has to examine the place and meaning of business in our societies. Even those who lead lives far removed from the factories and shops of manufacturing and commerce need to have a view on business, who it is for and what it is for. Directly or indirectly, their economic well-being depends on it. A recession brings home to everyone the importance of a healthy trading sector in the economy. When business goes down, everything goes down – jobs, tax receipts, house prices, government spending. Does this mean, however, that business is purely a wealth-creating instrument, best left alone to do what it has to do, or does it mean that, precisely because of its social impact, it has to recognise a wider accountability than making its owners seriously rich?

More directly, the business ethos has invaded our life. Everything is now thought of as a business of a sort. We are all 'in business' these days, be we doctor or priest, professor or charity-worker. *Every* organisation is, in practice, a business, because it is judged by its effectiveness in turning inputs into outputs for its customers or clients, and is judged in competition against its peers. The only difference is that the 'social businesses' do not distribute their surpluses. Americans, I was once assured, have always known this, but the hard reality of it had escaped most Europeans until lately.

Britain, however, has recently turned its schools, hospitals and medical practices, and even the service-delivery parts of the government, into independent businesses, funded still by the state but judged by their effective use of resources and required to compete for customers. When the full implications sink in of what it means to be 'businesslike', we shall realise what a revolution it will be to our way of life. One of the implications is, however, that all these organisations, in which one-third of our people work, will all have to answer the same difficult questions – what is this business for, and to whom does it belong? Are we who work in these businesses, be they social or commercial, their instruments or something more than that? What are our rights, and what, conversely, our responsibilities?

We have managed to evade most of these questions while we were preoccupied with our common enemy in communism. Anything, we assumed, must be better than such a centrally controlled system. This conveniently ignored the fact that many of our largest organisations were run in a similar totalitarian way. Now that communism has been discredited, capitalism must be its own sternest critic. Anglo-Saxon capitalism, when we see it exported in all its nakedness to the old socialist countries of Eastern Europe, is revealed to be good for some but not obviously for the many. It is, also, increasingly clear that there is more than one variant of capitalism. Michel Albert has spelt out the differences between the Anglo-Saxon version and the continental European version in his book *Capitalism Against Capitalism*. But there is also the capitalism of Asia, what might be called Confucian capitalism and, in particular, its Japanese variant.

The Anglo-Saxons have much to learn from the other varieties. Some hoped that the businesses of Britain and America would start to emulate the way in which the Japanese and the continental Europeans had a form of Chinese contract with their six different stakeholders – their financiers, employees and suppliers most obviously, but also their customers, their environment and society as a whole. Such a six-sided or hexagon contract inevitably changes the priorities of the business, leaving more room for the concerns of the other parties. A business, then, is no longer just an economic instrument. Ironically, the pressures of competing in a global world are pulling the others towards the Anglo-American model as fast as they, on their part, move towards them.

The question, 'What is a business for?' is addressed in the chapter on **The Corporate Contract**, in which I argue that the different systems of capitalism do indeed need to move together, borrowing the best of each other's traditions, in order to forge a new image of capitalism, one more obviously in the service of its society, but one still flexible and efficient.

The other question, 'To whom does a business belong?' is equally tendentious. I question whether the idea of a company as a piece of property which can be owned by anyone with enough money to pay for it, or bits of it, a property which can be bought and sold over the heads of all those who work and live there, is still a valid concept in an age when people not things are the real assets. Property is certainly not a valid concept when we think of the new 'social businesses'. So what sort of institutions are they? Instead of arguing who the rightful owners ought to be, I suggest a third angle – that ownership is not a valid or relevant concept, any more than property is. We ought instead to think of 'membership'.

The consequences of this line of thought are explored in the chapter on **The Membership Business**. Membership gives meaning, and responsibility, to those who work in the business. They cease to be instruments or employees and become enfranchised. Ironically, if we return to the old meaning of the 'company' we realise that a company was a group of companions, members one of another. That original meaning of a company still lingers on in the occasional theatrical 'company', or in some of the old livery companies in London, now charities not businesses. Perhaps we should rediscover the original meaning of the word.

The concept of a 'company' in this sense already exists in the way many volunteer groups and not-for-profit organisations think of themselves. Just as these organisations are becoming more 'businesslike', so we may see businesses looking to the non-profit arena for new models for themselves. The non-profit world understands all too well the combination of core funding and optional space, the doughnut principle, and is familiar with the necessity of Chinese contracts in a good cause. These organisations may, unexpectedly, hold the clues to the second curve of capitalism.

There are more clues in places as far apart as Michigan and Brazil. We need many more if capitalism is to prove that it has a human face.

8 The Corporate Contract

Capitalism has, supposedly, triumphed. Some claim there is no better way to run our societies than a mix of liberal democracy and free-market enterprise. Business gives wealth and opportunity to us all. Consultants and economists from the West swarm into the new market economies of central Europe, with their overnight cases, to show them how to do it our way in 24 hours flat.

Capitalism Triumphant?

The first results of the new capitalism are, however, far from reassuring. Industrial output in Poland fell by 35 per cent between 1989 and 1991; inflation reached 260 per cent. In Hungary, arguably the best-prepared of the new economies for the transition, food and basic expenses absorbed 45 per cent of the average household's expenses in 1989 but 70 per cent in 1991. In the Czech Republic they hoped, at best, to keep the fall in real wages to 12 per cent in 1991 and 10 per cent in 1992. Russia is a catastrophe, where the figures do not even make sense.

The full sad saga is forcefully described by William Keegan in his book *The Spectre of Capitalism*, where he comments:

> The terrible thing about the *sudden* adoption of capitalism is that the two necessary conditions preached by the reformers are in conflict. 'Price Liberalisation', needed to make the markets work . . . almost inevitably involves a disturbing

> acceleration in inflation as people rush to protect themselves from higher prices. 'Stabilisation' is, then, an uphill task in the face of 'liberalisation'.

The paradox is hard to balance. It was a director of West Germany's Audi who said, 'There are lots of books on how to move from capitalism to socialism, but none on how to do it the other way round.' He added, ruefully, 'We seem to be doing the research for that new book!'

Western capitalism in countries like Russia has come to mean 'trading'. The black market and mafias abound. 'Yellow Page' services proliferate. Street bazaars and back-yard markets are everywhere. Keegan reports that people can buy cars in Poland and, 48 hours later, sell them in Moscow for profits equivalent to 10 years of a professor's salary. Visitors happily spend the annual salary of a Russian on a piece of fashion or a call-girl. Such 'Wild East' capitalism, he says, is never going to be the foundation of a proper market economy. Nor does it, at present, seem to promise the sort of freedom to shape our lives that many hoped for.

The big manufacturing enterprises are lacking, but a country as large as Russia cannot survive on services and trades alone, nor will it be content to be the cheap labour shop for Europe after its past as a world power. 'We used to build rockets to circumnavigate the moon in this plant,' said one Russian colonel. 'Are we to turn round and make pots and pans to compete with central Asia?' Even in Hungary, only 10 per cent of the larger enterprises have been privatised. Most of the rest are probably not viable on their own.

Keegan was writing about Eastern Europe. My fear is that he could have been describing a possible scenario for Britain and America in a few years' time. The version of capitalism so triumphantly carried to those countries is the Anglo-American version. There are other versions, notably those of Japan and continental Europe, which have had a better record of combining liberalism and stability. They have made a better job of balancing economic freedom with relative equality, of giving more chances to more people.

The different versions of capitalism share certain fundamentals – free markets, the private ownership of assets, private direction of investment. They also share the idea of the hexagon contract. In each version the company operates in a space bounded by six interest groups – the shareholders or financiers, the employees, the customers, the suppliers, society and, lastly, the surrounding community and environment community. Where they differ is in the emphasis which they give to each of the interest groups. The difference is highlighted by the answer each version would give to the question 'What is a business for?'

What IS a Business for?

In my American business school in the Sixties the answer was clear. It was inscribed above the blackboard in every class so that we could not ignore it – 'maximize the medium-term earnings per share'. Medium-term, mark you, not short-term, and 'maximize' not optimise. Twenty-five years later things had not changed. Just before announcing his resignation as chief executive of IBM, John Akers complained that, 'The average IBM'er has lost sight of the reasons for his company's existence. IBM exists to provide a return on invested capital to the stockholders.'

From this basic premise all else flowed, given, of course, a perfect market and an intelligent one, managers who were clever, energetic and wise, and an educational system which provided an intelligent and rational workforce. Looking back, it is amazing that none of us challenged either the premise or the assumptions. Yet my own life up to then should have given it the lie. I had been the lowly regional manager in a distant outpost of a great oil company. I suppose that I must have seen the published results of the company but its earnings per share, its profitability, did not keep me awake at night, nor get me leaping out of bed in the morning. I was not a fool. I knew that any new project, rationally, needed to earn a rate of return above a certain figure, and my proposed projects were always expected to do just that, although neither I nor, as far as I know, anyone else ever checked whether those projects in fact lived up to their estimates.

If I'm honest, it was not the shareholders but my own self-respect which drove me. Sitting in that far-off country, the idea of maximum earnings per share was very remote, very intellectual, very unreal. I had, I was sure, a much more serious social function, as I told a maiden great-aunt back in Ireland who had complained that I was the first of the family to go into 'trade'. I was there to help produce things for people which were badly needed, in good condition, at a fair price, on time, without mucking up the local scenery or upsetting the local councillors or villagers among whom we lived and worked. It was a form of social contract, but, of course, it needed profits to make it work and go on working.

My business school in America was wrong, I am now convinced. The principal purpose of a company is not to make a profit, full stop. It is to make a profit in order to continue to do things or make things, and to do so ever better and more abundantly. To say that profit is a means to other ends and not an end in itself is not a semantic quibble, it is a serious moral point. A requirement is not a purpose. In everyday life those who make the means into ends are usually called neurotic or obsessive. We have to eat to live, but if we live to eat we become distorted in more senses than one. In ethics, to mistake the means for the ends is to be turned in on oneself, one of the worst of sins, said Saint Augustine.

'Profits are the principal yardstick,' stated the Watkinson Report on the responsibilities of the British public company 20 years ago, but a yardstick for what? And how can a yardstick be a purpose? It's like saying that you play cricket to get a good batting average. It's the wrong way round. You need a good average to keep on playing and to get into the first team. We need to clean up our logic.

Different Cultures, Different Dreams

Lester Thurow, in his book *Head to Head*, argues that Anglo-Saxon economics stem from the Anglo-Saxon emphasis on the individual and, in particular, on the individual as consumer. The individual is not so much interested in the work itself as in the results which that work will produce for himself or herself. Personal wealth is the result which the Anglo-Saxon wants, because that wealth will make possible the life style for which he dreams. The work is a means to an end, not an end in itself.

Take, for instance, William Caxton, who brought the printing press to England in 1477, an early example of technological pioneering: 'Caxton was an early and prominent example of a well-known modern type,' says Anthony Glyn, 'the individualistic Englishman following out his own hobbies . . . As a successful merchant he made enough money during thirty years to devote his later life to the literary pursuits he loved.' British businessmen, when pressed for their real purpose in life, nearly always say that they want to make their pile and then do something 'which really interests' them. Business is only a means to an end.

The British businessman tends not to be interested in sustained continuity. Private businesses, for instance, seldom turn into third- or fourth-generation family businesses. They are sold or go public long before then.

Many a British entrepreneur would feel that to ask children of the next generation to take on the business would be to constrain their freedom. The Victorian entrepreneurs who built Britain's industrial fortunes wanted their children to have nothing to do with business, but to be country gentlemen.

It is different in Japan. Thurow describes the Japanese business leaders as empire-builders and social builders, gaining their satisfaction from being part of a great and growing empire. To such people the use and ownership of production goods may be more important than consumption goods; they would, in fact, be happy to trade personal consumption for the success of 'their' empire. Imperial Rome, he points out, had many more grand public buildings than fancy private homes. In America it is, often, the other way round.

Japanese workers join a firm in much the same way as volunteers join an army, not for personal wealth or glory, but to be part of some great endeavour. Today it is the business enterprise which offers the best chance for empire-building. Given those attitudes it is hardly surprising that the Japanese put long-term growth above short- or even medium-term profits, indeed that profitability calculations hardly figure in some of their strategic decisions. To keep IBM at bay, Fujitsu won the computer contract for the water-distribution system of Hiroshima City with a bid of just one yen. The required rate of return for a 10-year R & D project averages 8.7 per cent in Japan compared with 20.3 per cent in the US and 23.7 per cent in the UK. As a result, there is more investment in the future in Japan than in the other countries. In 1992 Japan invested the equivalent of 34.2 per cent of its GDP in fixed assets. The figure was 16 per cent in the UK and 14.8 per cent in the US.

Germany is different again. Germany thinks of itself as having a 'social market' economy and not just a 'market' economy. Business is seen as serving all the people, not just its shareholders or even its employees. Heinrich Henzler, the chairman of McKinsey's German offices, has written that: 'Laws on co-determination, combined with a tradition of patriarchal concern, have made European C.E.O.s deeply committed to their employees, treating them more like partners in a long-term enterprise than anonymous "factors of production".' When he says European he means continental Europe, not Britain. He goes on to argue that this is a source of great competitive advantage.

Every employer in Germany of any size regards it as part of their duty to take part in the 'dual system' of workplace training, even though they may not employ the trainees at the end. They see that training as their investment in the continuity of German business, of which they will be a continuing part. The *Mittelstand*, the family businesses which are the backbone of the German economy, rarely sell out to others but are seen as a trust to be carried on by the family.

One reason for the small size of the German stock-market (only 665 stocks are quoted compared with 2,300 in the smaller economy of the UK) is that the pension provisions of these smaller firms are unfunded. The pension money is held in the company. The assumption is that the firms will always be around and be able to pay the pensions of their ex-workers. It also assumes that those workers would naturally want to work for the same company all their lives. Continuity is built into the system along with an acceptance of expensive social welfare policies, designed to take good care of those who are temporarily outside the system. German business exists for the good of all.

It helps, of course, that the firms are allowed tax relief on the reserves which they build up to pay those pensions, but they have the choice as to how they use those reserves in the meantime, unlike the separately funded pension schemes of British and American firms, where the moneys are managed by outsiders, charged with considering only the interests of the pensioners. Not unnaturally, the German firms often use those reserve funds to reinforce their links with key suppliers or agents by investing some of the reserves in their businesses, just as the Japanese do with *their* unfunded pension reserves. It is another force for continuity.

The role of the banks reinforces the sense of continuity. The banks are not short-term financial helpers, concerned mainly to make sure that their money is secure so that they can call it back to lend to others who are a better or more profitable risk. In Germany the banks are there for the long haul, with a stake in the business. In 1987 the *Economist* calculated that the large banks owned 10–25 per cent of the shares in 48 of the 100 largest firms, 25–50 per cent of the shares in 43 others and over 50 per cent of nine. In other words, every major firm was locked into the big-bank network and vice versa. No wonder that contested takeovers are almost unknown in Germany. They would not succeed.

The New Blend

Our versions of capitalism are the products of our histories. As a German Foreign Minister once said, 'The British were very generous after the war, they insisted on federalism, co-determination and single plant unions for us but took none of these for themselves!' Thurow and Keegan are not alone in seeing problems with the Anglo-American version, with its hint of selfishness, and favouring the German model, accepting that the Japanese version is probably unique to their culture. The Chinese, for instance, with their history of family enterprises are closer to the Italians and the Germans than their Japanese neighbours.

Paradoxically, however, although the German and Japanese models have been clearly the most successful in building rich and relatively equal societies, there are signs that as the world becomes one market-place, the versatility of the Western-style capital markets and the freedom of the individual in the Anglo-Saxon cultures become seductive. The third generation of the *Mittelstand* families are not as keen as their forebears on the idea of a family trust, if it locks them into one firm and one town for life. Pensions will soon be funded and that money will boost the German stock-market – to thrice its size in 10 years, some think. Meantime, as both the Germans and the Japanese acquire foreign shareholders in their pursuit of global empires, they are meeting with investors who cannot be expected to share the Japanese quest for economic supremacy, but want shorter-term rewards.

As a result, the balance of forces among the six interest groups is changing in all the countries. In Anglo-Saxon capitalism the shareholders have, traditionally, come first, with the other parties seen as a constraint, legitimate maybe, but still a constraint on the primary purpose. It is now accepted that all the so-called 'stakeholders' matter. The principle of the hexagon contract is now written into most corporate statements of purpose, even if the shareholder is still the first, because the most essential. The shareholder has to be the core of the corporate doughnut, but it is widely agreed that the business is not fully developed unless the interests of the other stakeholders fill the empty space in the doughnut.

In Japan the usual view is that the employees come top of the list, but Akio Morita of Sony maintains that it is really the customer who heads the others, not from any idealistic notion of wanting to please the man or woman in the street, but because the customer represents the empire they are seeking to build. Morita is now sounding cautious, because he senses a backlash from competitor countries who resent the competitive advantage which Japanese companies obtain by starving the other stakeholders in order to keep prices low for the customer. A readjustment is needed, partly for the sake of global harmony but partly, also, to placate the other stakeholders who would like bigger pickings.

In Germany there has always been a very conscious effort to balance the interests of the six stakeholders. Henzler calls it 'a social balancing act', arguing that business in his country has always accepted that homelessness, illiteracy and other social ills are not only morally unacceptable but are also economically harmful. Business has therefore been willing to bear the considerable social overhead because of its long-term benefits. Some rebalancing is now starting.

In the past, German firms refused to trade their stock on the New York Stock Exchange, arguing that the requirement to publish quarterly reports distorted the priorities of the business and distracted its management from its proper longer-term concerns. Recently, the need for funds to finance its restructuring and expansion has forced Daimler Benz to change its mind. Others will follow. Foreign stockholders will not share the preoccupation with Germany's social balancing act, any more than they want to encourage Japan's economic empire-building. German investors, too, want more than they have been getting. A recent survey of 11 stock-markets over the past 20 years ranked Germany number nine in terms of returns to investors. These investors are now growing restive. One group is even suing Deutsche Bank.

Even without this outside pressure, German business is worried lest the cost of the social balancing act may have grown too high. Jobs are draining out of Germany. BMW sites its new factory in the USA, Volkswagen is looking to Spain for its mega-plant. Hungary and the Czech Republic are close neighbours with skilled labour at a quarter the cost of German labour, even in the east of the country. The social costs of that eastern part are also stretching consensus to a breaking-point. One young German executive put it dramatically: 'If they had to buy some underdeveloped country,' she said bitterly, 'why couldn't they have chosen a smaller cheaper one?' The new generation of Germans may not be as prepared to pay the price for social cohesion as their parents were.

The Existential Company

As the cultures blend, the purposes of a business become less clear. Germany's social balancing act, Japan's economic imperialism, America's and Britain's priority on the returns to the owners, these all become more muted as the other forces in the hexagon contract become more powerful. What then is a company for in this new, more blended, world? The only real answer, I suggest, is 'for itself'. We might call it the existential company.

The existential company operates with the hexagon contract, but within the bounds of that contract is primarily concerned to grow and develop. Its continued existence, its immortality, is its purpose. It may, of course, turn out not to deserve immortality – the life-cycle of the average public company is only 40 years – but it is a worthy aim because, unless all the six interest groups are satisfied, the company will be unlikely to live that long. I liked the family business head who said, looking down at the roofs of the little Belgian town which was dominated, and employed, by his firm, 'We had to sit out two world wars, but they counted on us. In a family business you have to think beyond the grave.'

No one can lay claim to immortality. They have to deserve it. A company will only be allowed to survive as long as it is doing something useful, at a cost which people can afford, and it must generate enough funds for their continued growth and development. Existentialism in business is not, therefore, a form of selfishness. There has to be what James O'Toole, in America, has called stakeholder symmetry, and most of those stockholders are likely to have a vested interest in immortality. Employees, customers, suppliers and the community would all prefer that a business continued, as long as it was good. Even shareholders, now that so many of the institutions are locked into their stakes because they are too large to switch around, will settle for 'continuity provided it is justified by the results'.

'Stakeholder symmetry', however, doesn't get the blood beating any faster than 'shareholder value' which is why I prefer to settle for immortality.

Better not Bigger

What then, would be the purpose of such an existential company? The answer will be different for every business. Satisfying its financiers is a necessary condition, the core of the business doughnut, as is satisfying customers and stakeholders, but a necessary condition is not a purpose. That purpose may be, as in Japan, to conquer the world, but it can be less grandiose. You can grow without wanting to be the biggest or even big.

After one sun-drenched day in the wine country of Northern California I asked the owner of the winery about the future. He was passionate about his winery, he said; he was putting back every cent he could into its growth. 'Where can you grow?' I asked, looking around at the valley where every inch of land was by now fully planted with other people's vines. 'Oh, I don't want to expand,' he said, 'I want to grow better not bigger.'

Better not bigger. It is one definition of a purpose, one way to grow, one recipe for immortality. What we *are* can be as important an aspect of purpose as what we *do*. The existential hexagon company would, however, require some changes in the law, at least in Britain and America, because the rights of the shareholders would be severely curtailed. Perhaps not. The law in both countries already recognises the company as an entity in its own right. Lord Justice Evershed, summing up in 1947, said, 'Shareholders are not, in the eyes of the law, part owners of the undertaking. The undertaking is something different from the totality of its shareholdings.'

The judge was describing an existential company, one that exists in its own right, something which has a life and a future of its own. He was suggesting that all companies are, in law, existential. We have to take that judgement seriously, and give it meaning. We have to assume that every company has a life of its own which needs purpose and direction. It is an end in itself, not an instrument owned by others. If we don't, if there is no shared sense of identity to which all parties subscribe, there will be little chance of finding a compromise between the different interest groups. Each will then quite understandably fight their corner and the toughest requirement will become the dominant purpose.

To Find a Purpose

The Anglo-Saxon countries do not start with the cultural beliefs which still pervade the businesses of Japan and Germany. The leaders of business will have to create that purpose which commands assent. Essential though profitability is for the continued existence and growth of a business it begs the questions 'for whom?' and 'for what?'; it is not, in itself, enough. At present, to many people the answer to those questions seems to be 'the shareholders' and 'their enrichment'. The managers, with a proportion of their rewards linked to the share price, are seen as being allied to the shareholders rather than to the workers, unlike in the other countries. The workers and the other stakeholders in the hexagon are then seen by the managers as costs, and costs are things which, instinctively, we seek to reduce. There is seldom a shared sense of belonging.

In one week in the recession-afflicted Britain of 1993, four large public companies reported huge drops in profits, turning them into loss-makers in two instances, but they did not change their dividend. As the president of Britain's Board of Trade commented at the time, 'Presumably the implication is that shareholders can make more money by withdrawing their funds from the business than by allowing the business to invest in itself.' It does not say much for the hope of immortality in those companies.

Again, the figures speak for themselves: since 1975, British companies have retained, on average, 45 per cent of their profits for reinvestment, American firms 54 per cent, Japanese firms 63 per cent and German firms 67 per cent. In such a situation it made perfect sense, as it still does, for British shareholders to take their money out of generous British firms and invest it overseas where the companies clearly believe in their own long-term future.

Not all Anglo-Saxon companies think that way. Johnson and Johnson's credo is famous in America. Formulated four decades ago by President Robert Wood Johnson, it lists the corporate priorities:

– service to its customers comes first
– service to its employees and management comes second
– service to the community comes third
– service to its stockholders comes last

The credo was put to the test during the Tylenol affair when some bottles of its best-selling pain-relief tablets were tampered with, and several people died. J. and J. famously responded by pulling all 30 million capsules off the shelves.

In the long term they gained, because their reputation soared.

Johnson and Johnson might not be so renowned for their credo in America if it wasn't so unusual. It is no different from the batting order in any Japanese company. In a study by Fons Trompenaars, managers from different countries were asked whether they agreed that 'the bottom line' should not be the only real goal of a business, but that the other stakeholders should be taken into account. Ninety-six per cent of Japanese managers agreed with the statement, as did 86 per cent of the Germans, but only 53 per cent of the Americans. The British were in-between at 78 per cent in favour of the stakeholder balance.

If we don't change our ways more quickly, we may see capitalism in our lands deteriorate into the kind of Wild East now to be seen in central Europe, although at a higher level of consumption and corruption, no doubt. To think in terms of an existential company, striving for growth and immortality within the hexagon, is one handle on the problem. Another is to rethink what we mean by a 'company'.

9 The Membership Business

A business is owned by its shareholders. It is a strange type of ownership. To begin with, those owners normally have limited liability, something that goes with no other form of ownership that I can think of. Secondly, the 'thing' which they own mostly consists of people. Owning people, no matter how well you treat them, is considered wrong in every other part of life. There was once a time, in parts of Europe, where a man, in law, owned his wife. No one now, however anti-feminist, would think that right.

The reasons are to be found in history, but history, I have argued before, is not necessarily the best guide to the future. Limited liability was a most ingenious invention which allowed private businesses to take the risks which expansion required. It was a privilege given, then, a century and a half ago, to people who really did own their businesses, ran them and stood or fell by their success. They were locked into the fortunes of their enterprises. The 'property' they owned was physically there to see, bricks and mortar, machines, raw materials. The people were 'hands', employed to work the property, just as they used to be employed to work the land. It made sense, if they were to expand as fast as they might, that they should not have to put all their personal wealth at stake. Hence the privilege, granted to the people of a certain time and of a certain tradition. With the privilege came some implied responsibilities, for the welfare of the workers and the quality of the work. These responsibilities were not always honoured, but the privilege of ownership and limited liability endured. Without it the railways of Britain, for instance, would never have been built, nor would the industrial revolution have happened on the scale it did. Whether what was right then is right now must, however, be another question.

Owners or Punters?

Ownership may no longer be the appropriate concept, but if it is, then it is the proprietors of the private businesses who have the best claim to be the inheritors of that tradition, with its mixture of privilege and responsibility. Their futures are tied to the futures of the business. For the publicly owned businesses the situation is different. The 'owners' of these companies are, for the most part, institutions – investment funds, pension funds, insurance companies. They have no direct involvement with the business. They do not manage it or work in it. They do not know those who do. They are not locked in. The average shareholding by the big institutional investors in Britain is held for four years. Their responsibility is discardable. If things are not going well their best strategy is to walk away from the problem, to sell their shares. Fair enough. The rules allow it and their own shareholders or fundholders require it. The result is to turn the shareholders of public companies into what the *Economist* once called 'punters', equating them with the backers of racehorses at the track.

To expect the punters who backed the bay gelding to stay with that horse throughout its career, or to insist that the trainer took their advice, would not be reasonable. If they don't like its form they transfer their money to another nag. Punters or speculators they may be, owners in any real sense they cannot be. Devices to lock them in by tax incentives or legal requirements would be but 'sand in a free market' as these things were once described. Nevertheless these punters have an extraordinary privilege. They are, for the price of their bets, given a vote from time to time in the auction ring as to who should own their horse. This means that they have to be wooed, continually, for who knows when the auction bell may toll? Every public company, under these rules, is potentially up for sale every day.

It is argued that the constant possibility of the auction ring concentrates the mind of the trainer. It has been known to do that for the occasional bad corporation, but not always for its own good. I asked one supermarket chairman why he was expanding so energetically into France and Belgium, buying up competitors wherever possible. Was it, I said, to take advantage of the new enlarged European market? 'No,' he replied. 'We want to make ourselves so big and so complicated that no one will be tempted to swallow us up.' The best defence against being bought in that ring, apparently, is to buy. Yet all the evidence is that the bidder does worse, most times, than the loser at the end of the day. The cost is, presumably, judged to be reasonable if it gives one protection from the diversion of the auction ring, but it does nothing for the original business.

You do not even have to be under threat in the ring to feel the distraction. The chairman of one large German company was asked why he constantly refused to be listed on the New York Stock Exchange. 'Because,' he said, 'the requirement to report my results quarterly would distort the perspectives of my managers. The access to American funds is not worth that loss of perspective.' He may yet be forced to follow his compatriot companies into that auction ring, but he knows the risk. Managers and investors have different time perspectives, by the nature of their responsibilities. That is as it should be. What is needed is a compromise, not the dominance of pseudo-ownership.

Some say that making the managers, and perhaps also the workers, into the owners removes the pressure of that auction ring. But the history of management buy-outs in recent years suggests that owner-managers are just as susceptible to large offers as anyone else. I have known quite a few who profess a dedication to long-term stewardship in October only to be out to pasture, richer by several millions, in November.

Others look to create a consortium of institutions who will act as proper long-term owners – banks, pension funds, mutual funds and other companies – leaving other punters to flutter in the margin without affecting the long-term ownership. The pension funds, however, who own more than half of all British or American shares, are responsible for other people's money and have always shied away from locking themselves in. In America they are not allowed to sit on the boards of the companies in which they invest.

Some hope that the size of the funds involved will effectively lock the institutions into the index of stocks so that they will be content to stay where they are. There is little sign that those institutions, or, more precisely, their fund-managers, will be content to be so inactive as punters. And as for individual shareholders, one report predicted that the last individual shareholding in Wall Street would be sold in 2003. The idea that we could become a nation of small independent shareholders, which some dream of, is just that, a dream. Whether they would, in any case, behave any differently from their bigger brethren is open to doubt. Why should they?

There are a few signs that the punters are being pushed into behaving more like real owners. A batch of state legislation in the US has made hostile takeovers more difficult, forcing the shareholders to put pressure on the boards of corporations if they want change instead of waiting for someone else to buy them out. Several chairmen of major companies in the US and Britain have 'retired' rather more precipitately than they expected as a result of this pressure, but usually too late and leaving too much for their successor to do.

Property or Community?

Instead of fiddling with the rules, we ought to be asking whether we are still playing the same game as we once used to. Why is it sensible to think of an organised group of people as a piece of property, to be bought and sold according to its market price? Because that is what companies really are these days, organisations of people. A business does not have to be as rarefied as Microsoft to realise that its key assets are its 'human resources' and the kinds of intellectual advantage that they carry around with them – not just their creativity and their technical knowledge, but their network of contacts, their human skills and their experience. Everyone accepts that Japan's economic success has nothing to do with raw materials but is entirely based on the way they educate and manage their people. We have been slow to draw the obvious conclusion – that the same might have to be true for the rest of us; we must make our people our assets, and turn most of our property into the intellectual variety.

'Intellectual property' is a neat phrase, but it may delude us into thinking that the same traditions of ownership can continue. They can't. Intellectual property means people. Organisations are nothing if they are not communities of people, and a community is not a property. It does not make sense to say that a community is 'owned' by outsiders. A community is not a commodity to be bought or sold. A community has 'members' not 'employees' and it belongs to its members – only outsiders, not insiders, get to be 'employed' or hired by the community. If it needs money it raises loans or mortgages, secured, maybe, against part of the physical assets. It could, conceivably, sell a share in the future stream of net income – a form of equity – so that its financiers could share in its fortunes, but such a share would give no other rights. A community belongs to its members.

What would this mean in practice? Businesses would be self-governing communities. Limited liability would still apply, justified now once again because the business 'belongs' only to its members. Financiers would, in effect, hold mortgages but could only intervene managerially if the business reneged on its payments. Some mortgages would carry no repayment obligation but, instead, a share of the income stream for perpetuity. Mortgages could be traded, stock-markets would continue, but only as betting-rings not auction rings. Businesses would only merge or fold by decision of their members, who, doubtless, would normally take their financiers into their confidence. Outwardly, little would look different, but inside it would feel very different.

I spoke with the management of a smallish electronic company in East Anglia. They had had three different corporate owners in three years. It had not, they said, been very conducive to long-term planning or to morale. I could, at least, admire their British genius for understatement! They had become just part of the business portfolio of the big groups, to be bought, sold or exchanged as they rearranged their corporate-asset profile, to present a more pleasing picture to their owners in their turn. It did not make a lot of sense.

A View from Abroad

In Japan and Germany, for slightly different reasons, the idea of the company as a community, and of financiers as mortgage-holders, has long existed. Michel Albert calls it the Rhine Model, because it prevails in those countries which line the Rhine, but versions of it are found in Sweden and, with a slightly different twist, in Japan. It is, says Albert, who has worked with both Rhine model of capitalism and the Anglo-Saxon variety, markedly different from the property concept of a company.

In Japan the shareholders are more like preference debenture-holders; their dividends are related to the par value of the shares and not to the market value. Many of them are suppliers or associates of the business and get their rewards from being in business with a sound and growing company. They are the bankers, leasing companies, insurers, part-suppliers, distributors and agents who, as Carl Kester points out in a recent study, see their shareholding as the entry fee to a mutually beneficial system.

Unlike the Anglo-Saxon tradition, the board and management, in Japan, are not seen as the representatives of the financiers, but of the workers. The senior managers are not rewarded by linking them into shareholder interests by share options, as in Britain and America. Instead, they are linked, through a bonus system, to the performance of their workforce, the members of the company. By law, any merger or takeover requires the agreement of a majority of the directors of the company, but the directors are almost all insiders, career managers, representing the people with whom they work. If the financial returns are satisfactory the shareholder in Japan has almost no power.

Japanese companies will borrow fiercely to finance growth, but once secure will do their best to finance future growth out of retained earnings. In the 1980s, Japanese companies, on average, carried four times as much debt as American companies. Toyota, on the other hand, had no debt at all and was known as the Bank of Toyota because of its self-contained financial strength. Toyota does not want its investors to be its controllers.

The much-discussed lifetime employment policies of

Japanese business also fit the community concept. Members of a community cannot be expelled. They are there for life. The Japanese company will, however, make sure that they have as few people in their 'organising core' as possible and that they are the best around. It is not always realised that the lifetime system applies only to men, only to large organisations and only to full-time employees. It is generally thought that these true community members amount to less than 30 per cent of the total workforce. No wonder there is such competition to join one of those business communities. No wonder, either, that the organisations spend so much of their time on the learning and development of their people. They have no other choice. They can't sell their people-assets and buy in others.

In Germany and some of the other continental European countries the same concepts apply but for different historical reasons. Unlike Japan, German business is not dominated by the big names. In 1989 *Business Week* listed the 1,000 biggest businesses in the world. There were 353 American firms, 345 Japanese and 30 from Germany. Those are the only firms actively traded on the Frankfurt Stock Exchange, which anyway only has 665 stocks compared with the 2,400 in London's Exchange. Germany's strength, as we have noted before, lies in its *Mittelstand*, its small- to medium-sized family businesses.

Tom Peters, who first revealed the *Mittelstand* phenomenon to America, says that there are maybe 300,000 of these firms with anything from 10 to 3,000 employees. Less well-known, but equally important, are the family businesses of Northern Italy making knitwear, textiles, bricks, tiles, furniture, hydraulics, farm machinery – all the middle-technology, design-conscious products which are the staple ingredient of Italy's exports. The big industrial combines in Italy are mostly state-owned.

These German and Italian businesses are families. They want global reach but not global size. They concentrate on what they know that they can do well and make sure that it is good enough to be among the best in the world. That way they can grow better without growing bigger and can remain a family. Their financiers are investors rather than owners or controllers. They are the banks and insurance companies who are effectively locked in for the long term. It would be difficult to get rid of their shares except to another friend of the business.

The point of these businesses is to be able to go on doing it, profitably and enjoyably, for as long as possible. It is a way of life, not a means to an end. Since immortality is the point, and since shareholders are locked in and cannot be too greedy, the family heads inevitably think long-term, invest hugely in innovation and keep their core group small but excellent. These businesses, however, are families, not communities which belong to the members. The head of the family is still the owner. The best of them, and not all of them are best or even good, think of themselves as responsible for not only their children's futures but the futures of their workers' children. That way it makes sense to trade off the short term for the longer opportunities. Tom Peters records that the *Mittelstand* chiefs he met talked in decades not quarters, when, that is, they bothered to put a timescale on it at all.

Family businesses, however, depend on the family for immortality, and that tends to be a fragile base. The Italians talk of the 'Third Generation' syndrome, when the family talent peters out or goes in search of other pastures; rags to riches and back again, as the British put it. Many of the *Mittelstand* businesses are now approaching that third generation. The Sigmoid Curve is beginning to turn down for some. They are losing their innovative thrust, the family is becoming lazy, or greedy, or both. Some are looking for ways to sell. Immortality for the *Mittelstand* would be better ensured if the family came to mean the family of workers, the members of the work community. In the bigger companies the German concept of co-determination, which puts equal numbers of shareholders and workers on the supervisory board, is an attempt to create that sense of one family even in a large enterprise.

I suggested, in discussing the corporate contract, that companies could only keep the interest groups in balance if they were 'existential' in the sense that they felt completely responsible for their own destiny. This is only possible if they are independent, not owned or in thrall to outsiders. To the Japanese the company is a community. To the continental Europeans the best companies are run like families. Neither concept appeals to the British or the Americans. They both sound weaselly, undynamic. There is a word, however, an old Anglo-Saxon word with all the right history, until recently; it is a 'company', meaning a fellowship, a group of companions. Somewhere along the line it acquired its technical legal definition and lost its connotations. There was a time, maybe, when we had got the concept right.

The Reinvented 'Company'

The models of this old-style 'company' already exist in our societies in some unlikely places. We can borrow their ways but not, I think, their names. There is, for instance, the 'club'. A club is a place which belongs to its members and whose underlying purpose is its continued successful existence. It can best ensure that continuity by doing what it is best at doing. Its financiers are investors in its future, not owners or controllers, and its management works for the members not the financiers.

Perhaps, however, the most interesting models are to be found in the charitable and non-profit worlds. These organisations are owned by no one. They have constitutions, members, boards of trustees as well as boards of management, sources of finance rather than shareholders, and their purpose is their meaning. They are not properties, they cannot be bought or sold, although they can join forces, merge and make alliances. They have, in their doughnut, a core of professionals and, beyond it, a space full of helpers. These latter are often called 'associates', with limited rights of membership. The title which these organisations often carry is that of 'Society'. Thus there is, in Britain, for example, a host of Royal Societies for this and that, all prestigious, all communities. Society, or *société*, is the word in France for a business, and it might also serve in the Anglo-Saxon countries, but it would be preferable to reinvent the 'company' in its old meaning. The 'company' would have a core of 'companions' with 'associates' in the space around the core. It would be existential, responsible for its own destiny within the constraints of its hexagon, striving for immortality by doing better what it does best.

The Separation of Powers

A self-governing business, some will say, must be a licence for abuse. Self-determination has been a charter for scoundrels down the ages. Not all businesses deserve immortality. All this is true, but the market is a great corrective. Over time it sorts out the rotten apples in the system. That is not enough. Democracies, and federal democracies in particular, lay great store in the separation of powers. So it should be in the reinvented 'company'.

The legislative, or policy-making, function is separate from the executive, or management, and from the judicial, or monitoring roles. The roles overlap, in that the executive will propose most of the policies, while the laws which the judicial branch enforces are laid down by the legislature; but the functions are distinct and are usually performed by different people in different bodies. That practice is gradually being extended to organisations. It is seen most clearly in the charitable or non-profit bodies where the board or council is quite separate from the executive, and where, in Britain, there is an outside regulatory body, the Charity Commissioners, whose job it is to see that the charity is doing what it said it would. Continental European countries also favour two-tier boards for their corporations.

Britain and America are going the same way although they typically give the management board the title of 'executive committee'. The Cadbury Report in Britain in 1992, on the financial requirements of corporate governance, recommended that the roles of chairman and chief executive should normally be split and that there should be a substantial group of outside directors on the board. That is a small step in the same direction. More corporations are also putting their judicial function under a separate hat, with board committees to ensure compliance with their own rules and standards. There is even talk of a small trustee board of independent shareholders with limited powers of inspection and oversight, for the accounts and the appointment of board members.

After a fracas with Virgin Airlines in 1993, British Airways set up a new committee of its board on compliance. Some said it was a case of shutting the stable door after the horse had bolted, but it was, nevertheless, another step towards a proper balance in what is, in effect, a self-governing body. In Volkswagen in Germany and Phar Mor in America large-scale fraud was apparently conducted for years without the board being aware of it.

Effective and independent control systems are critical to the governance of self-governing bodies. The powers need to be even more visibly separated and separately staffed in an 'existential company'. It may be necessary to put the judicial or auditing powers outside the organisation, in the hands of an independent regulator. This already happens where the industry is dominated by a small number of companies who are, thereby, largely in control of their own destiny.

One existential community which, for many years, would have no truck with the separation of powers, believing instead that a concentration of the three functions of policy, execution and regulation would make it more effective, was Lloyd's of London, the insurance co-operative. They have changed things now, separating out the three roles into different bodies, but not before there had been a catalogue of frauds, mismanagement and bad policies, resulting in losses to its 'names', the members of this community, of some £5 billion over three years. There is a strong consensus that the concentration of powers led to a blindness to irregularities, to stupidity in strategy and laxity in management, rather than to any increased effectiveness.

The Membership Contract

In the end, however, a self-governing club in a competitive system should only survive as long as it deserves to survive. One threat to that survival will be a constantly changing membership. There will be no pressure to plan for the future of the children if even the fathers and mothers of those children are unlikely to be around in five years' time. The current tendency in Britain and America to use the organisation as a stepping-stone in a personal walk to glory and riches will make a nonsense of an immortal club. Loyalty has to be reciprocal. Temporary contracts will beget temporary time horizons.

We need to see more lifetime contracts again, remembering, however, that a corporate lifetime is going to be much shorter in future. I can see a period of apprenticeship, or articles, for new young employees, of perhaps five to seven years, followed by a fixed term 'tenure' contract of from 10 to 20 years, during which time they will be full 'members' of the 'company' or club. Professor Iwao has calculated that the average stay by core staff in one firm in Japan is 14 years. Lifetime in practice means only long-term. People might well serve their articles with one company and then join another as member, as accountants already do. After their membership period expires it could be renewed or they may well 'go portfolio', becoming independent advisers or suppliers to their old club. Membership will then be a privilege, akin to partnership but with limited liability, for a select minority – the core – requiring reciprocal obligations and loyalty.

In 1993 the Director-General of the BBC was discovered to have sold his services to the BBC through his own private company. This was a perfectly legal device which was common in the world of short-term contracts of commercial television from which he had come. It was not thought appropriate in the culture of the BBC where long-term loyalty was the norm, even though it is operating in the same industry. The feeling expressed by many was that this loyalty could hardly be demanded from others if the person at the head of the organisation appeared to see his own job as a temporary assignment.

It was a dispute symbolic of the times. More and more professionals think of themselves as on a temporary assignment with an organisation. Loyalty goes first to one's team or project, then to one's profession or discipline and only thirdly to the organisation where these skills are practised. In the City, whole teams of dealers and analysts move together from one organisation to another. Senior jobs in business are often on three-year contracts. Executives, like doctors, move from location to location as they advance their professional careers. 'Company' loyalty will be very short-term if this trend continues. The BBC was right to be alarmed.

My own belief, and hope, is that this trend too will be subject to the Sigmoid Curve, and that companies will want to hug their key 'members' to themselves for the duration of their, shorter, working lives. To do that they will need to give those key workers all the benefits of membership, including the effective rights of ownership. These rights will have to go beyond the current fashion for share options which gives the holder only a minority stake in the punters' bets, but will be something more akin to partnership, locking the key people into a self-governing membership group. Only then will we find the motivation to plan for immortality. Few would want to commit themselves to an organisation owned by punters.

Glimpses of the Future

If, however, the business as a self-governing, membership organisation is such a good idea, why don't we see more of them? The number, and the record, of co-operatives and their ilk is small and, with some notable exceptions, poor. Co-operatives often confuse ownership and management. Because ownership is in common they think that management also has to be shared. Democracy, however, does not require that all who vote should also have the right to manage, or even to demand a referendum on every decision. That way chaos lies. German and Japanese organisations do not make that mistake.

There are also, however, the ESOP companies, the employee share-ownership schemes which many companies have adopted as a way of giving their workers a stake in the organisation. The evidence on these is mixed. Some make a difference to motivation and commitment, most don't seem to matter much either way. A well researched study in America by Corey Rosen and others, found that the ones which did work had large employee contributions (8–10 per cent), a true philosophy of partnership and multiple ways of participating. The percentage of stock owned by employees made little difference, nor did the stock-market performance of the shares. In other words, it was membership not ownership that really mattered. If there was no sense of membership the ownership made no difference.

In Britain, the best-known membership business is the John Lewis Partnership, with its chain of retail stores. This business 'belongs' to its members, who receive dividends from its profits, elect their chairman but entrust the management to a conventional executive structure. They do not, however, own shares which they can sell. It is a true business community. It has had, however, few imitators.

This must be because we are hung up on ownership, on the idea of property. It is largely the fault of an outmoded legal system, in Britain at least. George Goyder, in his book *The Just Enterprise*, points to a prescient comment by Lord Eustace Percy in 1944:

> Here is the most urgent challenge to political invention ever offered to the jurist and the statesman. The human association which in fact produces and distributes wealth, the association of workmen, managers, technicians and directors, is not an association recognised by the law. The association which the law does recognise – the association of shareholders, creditors and directors – is incapable of production or distribution and is not expected by the law to perform these functions. We have to give law to the real association and to withdraw meaningless privilege from the imaginary one.

Our rules do not allow for a wealth-creating club which is not someone's piece of property. It does make it difficult to balance the contradictions inherent in a company owned by some and worked by others, but managed by the agents of the owners, because, whatever the rhetoric, it will be hard to find the elusive common purpose. Ultimately, we shall have to change the rules.

I take heart, however, from that research on share-ownership plans. Ownership makes little difference unless there is a sense of membership; therefore, presumably, the technical conditions of ownership make little difference if there is a real sense of membership. Laws tend to follow practice not lead it. If we can create that sense of membership in our organisations, by more subsidiarity, more twin citizenship, more sharing of the added value of closer teams and better boards, then, conceivably, the so-called owners would revert to their proper role, as financiers and owners only of last resort. The formal position would be irrelevant, as it is in Japan and, to a lesser extent, in Germany.

It is hard, nevertheless, to conceive of our giant multinationals and other mass organisations becoming membership 'companies'. We may well, however, see these organisations breaking down into alliances of much smaller ones. You no longer have to be big to be global in the information age. The *Economist* in London has 55 journalists but covers the world, in scope and in readership. The *Economist*, too, is owned, in effect, by one proprietor who holds the bulk of the voting shares. Other financiers have stakes with smaller voting rights. There is a board of management, to oversee the business, and a board of trustees to protect editorial freedom. Given its current benevolent proprietor, the *Economist* is almost a membership 'company'. It feels like one when you visit it. Give the voting shares to the members and the model for the 'company' would be complete.

Words and titles can help, even without the legal backing. When Ricardo Semler decided to make his Semco company more of a self-governing community he called the directors 'counsellors', the senior managers 'partners' and everyone else 'associates'. He then had to live up to the expectations the new words raised. He described the results in his book *Maverick*. If it worked as well as it did in the difficult conditions of the Brazilian economy, and without any change to the legal situation of ownership, it must be possible elsewhere. Ralph Sayer did the same at Johnsonville Foods, calling his workers 'members' and his managers 'co-ordinators' in order to symbolise the new order, one in which profit-sharing and autonomy went hand in hand. We need not wait for the law to change.

Redesigning Life

Changing the structures of our institutions, and changing the meaning of business, will help to clear a pathway through some of the paradoxes, but we can only find meaning and value in our lives by living those lives. A time of chaos is a time of opportunity. The old patterns are changing. We don't have to live as our parents lived. We can shape our own lives more easily than ever they did. The fragmentation of work looks scary but offers new freedoms at the same time. Time's paradoxes mean that the old blocks of time, the 40-hour week and the 50-year career, are probably gone for ever as a way of life. Women have more opportunities, as well as all the old responsibilities, responsibilities which can themselves be new opportunities for men, for the old divides are breaking down, slowly.

At the beginning of the century, 50 per cent of workers were independent. They were outside the formal organisation. By the end of it the percentage might be the same again, yet only 20 years ago 90 per cent of us worked inside the organisation. New thinking, aided by new information technology, is beginning to alter not only the way we work but where we work, when we work and how we charge for it. The result is a new way of thinking about the organisation of work and the way we use people to do it.

Governments are worried because work seems to be disappearing. It is. But it is also becoming individualised and informal, and therefore invisible and unregulated. Jobs are disappearing but not necessarily work, if we define work as useful activity. Pushing jobs back into organisations will be counter-productive. The more governments try to set proper terms and conditions of work, by, for instance, the Europeans' Social Chapter, the more they will encourage the spread of the minimalist organisation and further individualisation. Unfortunately, we are often our own worst employers, putting up with terms and conditions of work which no decent employers would contemplate inflicting on their people. We shall have to learn to live with this second curve of work. The first one will not return. However, this is our chance to make our working time more suited to our taste, to shape our own raincoat, if you will.

The demise of the traditional job, the re-chunking of time and the new areas of choice for would-be parents, combined with longer and healthier lifespans, mean that the traditional sequence of events in life – school, job, house, children, retirement – is no longer fixed. Flexilife is now the mode. Serial monogamy seems to be the new respectability. 'Has your wife got a daughter, too?' the woman asked, when I was telling her something about my daughter. The assumption was that at my stage in life I would be on my second marriage.

To add to the options, a whole new section of life has opened up beyond the time of full-time work or parenting. There will be some 25 years of healthy life for most of us beyond the job – a new compartment. But it will not be 'retirement'. We won't want to retire and most will not be able to afford it. Work will still go on, but it will be, mostly, work at a different pace and work of a different kind, but work it has to be, because work provides the structure of our lives, the core of our individual doughnuts.

The new divisions of work, time and life will redraw our maps of society. The new maps offer the chance of greater individual choice, but also of individual calamity. It will be more confusion at first; it already is, with people hankering after that first curve they had grown to know. In time, we shall learn to balance the contradictions and enjoy the second curve. It will be a new way of living and working for all of us. As always with the second curve, it will be the new generation who will find the adjustment easiest. The opportunity lies in the chance to live several different lives in the space of one – spliced lives, I term it. For their elders, the Third Age offers another chance to live some of the life they wish that they had had. They would be foolish to turn it down.

10 Working Time

Work is not what it was, inside or outside the organisation. The paradox of time combines with the changing nature of work, to force us to rethink what we mean by the whole idea of work and time – when we work, where we work, how we work and why we work.

The Minimalist Organisation

I was meeting with a journalist in Atlanta. We talked at her desk, cluttered as it was with papers, telephones, keyboard and screen, in the middle of the vast newsroom, 200 people chatting, typing, telephoning, smoking. There was no chair for me. I perched on her desk. 'It's not going to be easy,' I said, 'talking to you like this. Do you not work at home, at least for some of the time?' She smiled ruefully. 'Never. Of course I could do much of my stuff there, and I would do it much better with less hassle, less travel, less noise and I could always come in here when I needed to, but my work could all go down the telephone line – as it does anyway,' she added, 'from this desk.' 'So why don't you?' '*They* won't let me.' She pointed to the end of the room where, behind two large glass windows, sat the two deputy editors. 'They like to have me where they can see me and shout at me.'

One day those people will realise that an office, even a newsroom, does not have to be designed as a factory and that the cost of all that space in the centre of Atlanta is not worth the convenience of seeing and shouting. One day.

We all want our work clubs, places to go to, places to meet and greet and eat, but we no longer have to work there all the time. Unless, of course, you are there to serve the customers or the clients, in the store, in the reception, in the classroom, hospital or restaurant. These servers, however, do not have offices, private spaces of their own. Their work is where the client is. It is those offices with their private spaces which will gradually go, and with them a way of life. Many will lament their passing, but sentiment and nostalgia will not prevail in the new businesslike age.

We have to rethink the organisational contract, what we mean by an organisation, what we expect from it and what we are prepared to give to it. No longer can society rely on these new businesslike places, which are very unlike the businesses of old, to provide life and livelihood for everyone, even in times of economic boom, to collect their taxes from them or to pay their pensions. No longer will the office or the plant be the home-from-home for most men and many women. No longer will a career mean climbing the ladder of jobs in an organisation. For one thing, there will not be more than three or four rungs on the ladder. No longer can one expect to sell 100,000 hours of one's life to an organisation. No longer will your job title define you for life, or even for very much of it.

There has to be a major shift in the way we think about organisations. They are now living up to their name – they are *organisers* as much as or more than they are employers. They are minimalist. It shows up in the numbers – many fewer people inside the large organisations, more people in small ones, more people working on their own, more people, unfortunately, left without any work at all because they do not have the specialist skills needed both inside and outside the organisations. Boom or bust, it makes little difference. In Britain's boom years of 1985–90 manufacturing output went up by nearly 19 per cent, but the manufacturing labour force still fell by nearly 5 per cent. It simply falls faster in recession. Now it is the turn of the service sector, where productivity has hardly been tackled until recently. By 1993, in Britain, as we noted, only 55 per cent of all those in work or seeking work were in full-time jobs inside organisations. In the US it was 60 per cent. The figure in other countries is higher but coming down. More by accident than design, Britain leads the way. Before very long, having a proper job inside an organisation will be a minority occupation. Then the world really will have changed. What then will everyone do? How will they live? How, metaphorically, will they fill out those raincoats?

This is all happening at a time when the total workforce, those who are able and wanting to work, is growing. There may be some 25 per cent fewer teenagers now than there were 10 years ago, but we have to remember that this dip was preceded by a bulge, a bulge which is now in its thirties and will be around and looking for work for another 20 or more years. The workforce is getting older, certainly, but it is also growing, not least because the wives or ex-wives of many males in the bulge years now want to re-enter the workplace. The workforce in Britain will probably grow by one million during the Nineties and in the US by as much as 12 million. It is no longer, however, going to be, largely, an employee workforce. That simple fact is going to make all the difference. What was a way of life for most of us will have disappeared. Organisations will still be critically important in the world, but as organisers not employers.

From Diamonds to Mud

Inside the organisation it is also changing, as the principle of the doughnut begins to spread and organisations become redesigned and re-engineered. In times past, almost all the spaces in an organisation were predetermined. The prescribed core filled most of the jobs. In those days it was fashionable to make organisations as predictable as possible. They were designed like railway timetables, in the hope that all one then had to do was to press the button and watch the trains run on their preplanned routes. That way efficiency lay. Too much discretion lower down disrupted things, set off unpredictable chain reactions. No one would want the train-driver to be creative and miss out a station or two in order to improve the running-time.

Those were the days before chaos theory or the newer complexity theory in science. The Newtonian classical view of science pervaded even the pragmatic world of organisations. There should be an explanation and a set of rules for everything, everything should be planned and predicted in a properly ordered world. People were part of a well-ordered system. If only they did as they were supposed to do, everything would work fine. As people become more educated and more expensive, it does not make sense to treat them as automata, nor do those better-educated people enjoy having so little space for discretion in their jobs. The first, inner circle, of the doughnut has gradually become smaller, the outer one bigger. In-between, the space expands.

Ian Gibson, chief executive of Nissan Motor Manufacturing in Britain, told the graduating class of the London Business School that his first discipline was physics.

> I tend to think of things in a scientist's terms – in this case in terms of the difference between crystalline and amorphous structures. As an example, the most easy way to recognize a crystalline structure is in a diamond, and perhaps the most common, if unglamorous, amorphous material is mud. The typical western organization is crystalline; clearly defined, facets that have their own shape, with obvious joints between them. The features of our organizations are comparable – clear definitions of role and responsibilities; well-defined

> boundaries within different parts of the organization; each part of the organization in a known and fixed relationship to the others.
>
> In contrast, Japanese organizations are more like mud. They are far more blurred, separation between responsibilities and functions is ill-defined and in a constant state of flux. Reverting to the analogy, a diamond is clear, tough and precise. Mud is vague, it changes shape and form. It has, however, one over-riding benefit – it is easily shaped and changed and is flexible and responsive to external forces and circumstances. As organizations, we must become increasingly able to change quickly and easily. This means building on and around people's abilities rather than limiting them for the convenience of easily recognized roles.

For organisations the opportunity is now there to apply the doughnut principle to most of their work, devising a structure made up of muddy doughnuts, a system of interlocking double circles, in each of which the inner circle, the core, is tightly specified and controlled, as are the outer limits of authority, but where the space in the middle is to be developed. In a lax organisation, that can be a recipe for anarchy, if they cannot manage the inevitable contradictions. Individuals and groups may exploit their spaces to their own satisfaction, but not necessarily to the organisation's advantage. The centre cracks down, cores expand again, individuals resent the contraction of their discretionary space and mutual resentment saps morale. Large doughnuts work best when there is a clear consensus about the purpose of the work and the goals of the organisation.

More and more we shall see organisations divide their work into project teams, task-forces, small business units, clusters and work groups – smart words for doughnuts. These groups will change shape and membership as the needs of the organisation change. Individuals may well work for more than one group at the same time, for multiple membership is a feature of doughnut organisations, with one group having an operational responsibility, another an advisory role and a third having a temporary project assignment. It all helps to make life exciting, but much less predictable. Organisations, for example, are no longer guaranteeing to provide planned careers but are instead offering 'career opportunities'.

Look at an advertising agency for an approximate model of how we shall all be working tomorrow. An advertising agency arranges its people in clusters and task-groups, two sorts of doughnut. The clusters are clusters of expertise, the creatives or the planners or the people who book the space in the media. They are drawn from these clusters into a range of account groups where they work on the requirements of a particular client or product. They may work on several different account groups, and the membership of the groups will flex with the demands of the work. It is a fluid-matrix organisation. So is any consultancy firm. So is a hospital ward or a surgery team – they are all, conceptually, doughnuts, with their members drawn from other specialist doughnuts. They all have a tightly specified core but a space for initiative and improvement, often a great deal of space. The better the people, the bigger the space can be.

The old language of management no longer seems appropriate. It never was appropriate in some quarters. Professional organisations, doctors, architects, lawyers, academics have never used the word manager, except to apply it to the more routine service functions – office-manager, catering-manager. The reason was not just a perverse snobbery but an instinctive recognition that professionals have always worked on the principle of the doughnut. This was necessary because every assignment was slightly different; flexibility and discretion had to be built in. It was possible because the requirements of the profession, its rules and disciplines, meant that one could be reasonably sure that whatever one's colleagues (note the word) did in the space between the circles would be acceptable. Doughnuts work well when everyone knows, not only what the purpose is, but what the standards are.

The Portfolio World

Outside the organisation, things seem even more confused for that 40 per cent or so of outsiders, most of them reluctant independents. They may be the way of the future but few wanted to be the pathfinders. There are, however, some trends and ideas beginning to emerge. As organisations restructure themselves into their minimalist shapes, they are going back to buying produce not time. That amounts to taking the price-tag off time. Instead of employing their lawyer they now, in the minimalist organisation, buy his or her services, for a fee. Professionals and craftspeople have always charged fees. The fee will include a time element, but it will also charge for the quality of the work, for reputation and reliability. When Ruskin sued the artist Whistler for overcharging him for a portrait which, Ruskin claimed, he had dashed off in a hurry, the judge asked Whistler how long it had taken him to paint the portrait. 'Ten minutes,' Whistler replied, 'and a lifetime of experience.' It is the quality of the produce one is paying for, not the time spent on it.

More and more individuals are behaving as professionals always have, charging fees not wages. They find that they are 'going portfolio', or 'going plural'. 'Going portfolio', I suggested earlier, means exchanging full-time employment for independence. The portfolio is a collection of different bits and pieces of work for different clients. The word 'job' now means a client. My wife, a portrait-photographer, has, at any one time, a number of 'jobs' on the go, as does our builder when he tells us why he cannot fix the roof this week. I told my children, when they were leaving education, that they would be well advised to look for customers not bosses. If they could find people who were willing to pay them money for something they could make or do, that would be the best qualification for impressing a boss when they did finally want to join an organisation and sell their time to someone else. They have 'gone portfolio' out of choice, for a time. Others are forced into it, when they get pushed outside by their organisation. If they are lucky, their old organisation will be the first client in their new portfolio.

The important difference is that the price-tag now goes on their produce, not their time. I read, enviously, of the man who had met a man in a nightclub in London. They got talking. The second man was from the Middle East, looking for a solution to a big and tricky problem of irrigation in his part of the world. It just happened that the first man knew a woman who ran a business which turned out to have exactly what was needed. He received a £5 million commission for the introduction. It was not time that he was paid for but a crucial bit of knowledge, rather like Whistler. We can't all be so lucky, or so well connected, but in smaller ways the principle always applies; the price-tag is on the produce not the time.

It works the other way round, too. If I was paid by the hour at the average national wage for writing this book I would be counting my income in many tens of thousands of pounds. Sadly, the royalty advance takes no account of my time, but prices my produce at the level my publisher thinks it will fetch in the market-place. I therefore sell my time cheaply to myself, in the hope that it will be an investment worth the making. That way, I know, many authors starve! The reality, however, is that it is applied intelligence now, not time, which is the crucial element.

Professionals, the knowledge-workers of all types, are obvious candidates for portfolio lives. So are those who make or fix things, the traditional fixers and makers like plumbers, builders, carpenters and electricians, but now, also, the new fixers: the agents, brokers, conference-organisers, house-finders and sellers, travel agents and tour-arrangers. There are the new servant businesses, often one or two partners with a supporting cast of occasional stringers: the cooks, drivers, gardeners, health specialists, language-teachers, child- and house- and dog-minders, cleaners, even, I am told, people you can pay to change your light bulbs. There are, also, the old and the new crafts: the potters, weavers, bakers, painters, writers, computer software-designers and photographers.

Read the yellow pages in any city, anywhere, to find the portfolio world. These people charge for their produce not their time. It is not that they don't work as long or as hard as employees, because most of them work both harder and longer. The difference is that they have more freedom to chunk their time in different ways, if they so choose. What matters now is *how* we use our time, not *how much* of that time we use. When you can use time in this way it is a form of freedom. Those who charge by the hour can only make more money by working more hours. Those who charge for their produce can get richer by working smarter, not longer. This has always been true for some activities. It is now potentially true for almost everything. A wise jobbing gardener will quote you for keeping your garden in good trim, not for so many hours a week, and if you are wise you will accept this basis because the onus is now on him or her to use their time productively, not on you to make sure that they do.

Technology enables more and more people to go portfolio. Organisations are latching on to the possibilities, as we have seen. They are even extending the principle of product not time to their own internal operations. To a unit, a group or a person they say, increasingly, as the farmer or the businessman once said in times gone by, 'Do this by this date; how you do it is up to you, but get it done on time and up to standard.' This is subsidiarity in practice. What it means is that, even inside the organisation, people have more discretion as to how they chunk their time. If they want to chunk it in fewer bigger segments, they are, in theory, free to do so.

As this practice spreads, the difference between full- and part-time work will be mainly one, not of time, but of rights and entitlement. I have little doubt that we shall, increasingly, see both laws and best practice equalising the benefits, proportionately, between full- and part-timers. This will happen, politically and legally, because part-timers are no longer marginal extras in the workplace and will demand more social justice. We can see the Social Chapter in the Maastricht Treaty as one piece of evidence of this in Europe. It will happen, anyway, in the forward-thinking workplaces because they will want to bind their growing 'peripheral' workforce to themselves. Left neglected on the periphery the part-time workers will lack loyalty and will leave the business dangerously exposed.

Juliet Schor wants those organisations who still buy time rather than produce to specify how much time they are buying and to offer to repay extra time with time. Everyone would, then, in effect be on an annual-hours contract, or some variation of it. Some prized and special people might sign on for 3,000 hours, or 60 hours a week and two weeks' holiday, others for the more conventional 2,000 hours or less. People would have the right to claim repayment of the extra hours worked during the year by doing fewer hours the next year. Obviously the sums, and the repayments, could also be calculated on a monthly or six-monthly basis if so desired. Overtime would be repaid by undertime.

Patricia Hewitt describes the principle as time-banking. If you build up time credits in one part of the year, or over a number of years, you can draw on them later. You could do it, practically, by putting the money equivalent to a set number of hours per week into a pension fund, so that it would, under existing legislation, be tax-free. You then draw out that money when enough has accumulated, and use it as a substitute for your wages for a period. There is a proposal in Norway to allow people to do it the other way round, to exchange future earnings for present time, like running a loan account. If I wanted to take six months off to help with my family, I could pay it back later by putting in the normal number of hours but only drawing the pay for, perhaps, 80 per cent of them until the loan of time was repaid. My own preference would be to find a way to pay for time with time and to keep the money out of it but the practicalities are, I accept, difficult.

Women's Working Time

The concept of time-banking, whichever way it is implemented, would help to solve a growing dilemma for organisations and for many women. As organisations go minimalist they shrink their cores, pursuing that elusive formula of $\frac{1}{2} \times 2 \times 3$. Those in that core are, then, better-paid but harder-worked. The 3,000-hour year and the 60-hour week are not uncommon. Organisations are greedy places when they have bought you for a year or more. For these jobs the organisations want quality people, well educated, well skilled and adaptable. They also want people who can juggle with several tasks and assignments at one time, who are more interested in making things happen than in what title or office they hold, more concerned with power and influence than status. They want people who value instinct and intuition as well as analysis and rationality, who can be tough but also tender, focused but friendly, people who can cope with these necessary contradictions. They want, therefore, as many women as they can get.

They want them, or should want them, because not only do women make up at least half of all the well-educated people in our societies, but they are also more likely to exhibit the kind of qualities listed above. Men have these qualities, too, of course, but male conditioning over the generations, in both home and workplace, has emphasised singularity of purpose, one thing at a time, rank and formal authority, toughness rather than tenderness, rationality not intuition. Women, on the other hand, over the generations, have had to make things happen and get things done, with or without formal authority. They have had to handle endless variety in managing the home, taking decisions on inadequate information, backing hunch and judgement. They have had to be in turns disciplinarian and loving mother. Not all women do these things well, by any means, but few men have had that much practice.

Organisations need talented women in their core jobs, therefore, not only for reasons of social fairness, important though that is, but because many of those women will have the kinds of attitudes and attributes that the new flat flexible organisations need. If they screen out the women they will handicap their futures. Yet screening out is exactly what they are in danger of doing. It is difficult to combine those 60-hour jobs with raising a family. Many don't try. A 1992 survey by Britain's Institute of Management found that 86 per cent of the married men in their sample of executives had children, but only 49 per cent of the married women.

Every working woman, they say, needs a wife, particularly if she has children, and some are fortunate in having a partner who will be house-husband and full-time parent. Others delegate these roles to nannies, home-helps and crèches. Many very competent women do not want to do either, or cannot. For them, the portfolio life with its, now, expanding possibilities offers the flexibility and freedom to combine work with home. They would often, however, have preferred to stay in the core in some way, or at least to have the chance of going back when the children are a little older. Organisations can ill afford to lose them. Time-banking, and the organisation re-engineered to fit the process not the function, creates some of the flexibility which these women need. The chance to deliver the work in their own way without having to conform to rigid timetables, combined with the right to bank or chunk time so that school holidays, for instance, can be mostly time away from the office, would allow more women to keep a role alive in the core while the kids are young.

Other things would help. Organisations could rearrange their time so that all normal meetings took place on fixed days of the week, Thursdays and Fridays, perhaps. This would allow people to be physically absent from the organisation for part of the week without missing the meetings which are still the warp and woof of organisational life. One day we shall have to learn how to manage our dispersed organisations without meetings, but until that distant day comes we should at least take pains not to marginalise any absent friends. We can always be in touch telephonically and electronically, of course, but those who work four days a week say how strange it is that the really important meetings always seem to take place on the fifth day! Only in real emergencies, after all, do organisations now have meetings on Sundays. If they can rearrange time to exclude Sundays they have already granted the principle; now they need to extend it.

As we have noted, organisations are changing parts of their offices into work clubs. The 'outworkers' will be free to use the clubhouse at any time and will have their own personal 'cart' or 'hot desk' or 'puppy' which they can pull out, plug in and key in, giving them immediate access to all the systems and an instant extension number when anyone calls. IBM and other electronically minded organisations already do this. So does any small television production company where office space is at a premium and where most people are out most of the time.

Regular meetings in these 'clubs' will be scheduled for the 'meeting-days' but the rest of the time there will be no way of knowing where anyone is or what they are doing without contacting them directly by phone, fax, E-mail or voice mail and asking them. When that is happening to the majority of executives – as it will – it will no longer make sense to think of people as full-time or part-time. They will be paid for the type and size of the job they are required to do, an internal fee. Their time is now theirs to manage more than ever before. They can use it wastefully or thriftily; either way their pay will be the same. We will have put the price on the produce not the time. Time will have its new and flexible compartments.

Rethinking working time does not do away with the difficulties. It makes those difficulties more manageable. It puts more responsibility on to the individual in return for more independence. To make the balance fair, organisations will need to be less greedy in their expectations of the people they buy by the year and less mean in dealing with those they buy by the hour. They will undoubtedly need the encouragement of the law to do the latter, because employers like their labour cheap, although all the evidence is that you get more net added value in a knowledge economy from better dearer workers than cheaper worse ones, as long as you give them the chance to add more of that value.

11 Spliced Lives

'Live as if you will die tomorrow, but plan to live forever.' It is a useful maxim because perceptions change with age. We look as far ahead as we look back. I used to be frustrated because I could not bribe my tiny children with promises of treats the next weekend if they would only keep quiet that afternoon. They could not connect the two events. Slowly I realised that a week in the life of a small child is, maybe, 20 per cent of all that they can remember. It is like asking a 60-year-old to take account of something 15 years ahead when deciding what to do today. Young people live in the present. That is part of their delight. It is also part of their problem because, in this complex world, it seems harder and harder to find the kind of balance we want in any one bit of it. It need not be so. We can now flex the four traditional stages of work and four different types of work. We can, if we so want, put them in a variety of combinations, splicing a life together by twisting its strands in the way we want.

The New Flexibility

In the old Hindu scriptures, life had four stages: student, householder, retirement and sannyasin, that final stage when one neither hates nor loves anything. Nothing about work or jobs in those times! Shakespeare added infant, lover and soldier to give us his seven ages. Gail Sheehy, writing her best-selling *Passages* in the 1970s, settled for four, this time divided by decades: the Trying Twenties, the Catch Thirties, the Forlorn Forties and the Refreshed (or Resigned) Fifties. Not a happy-sounding list, nor a very long life! Daniel Levinson, who wrote *The Seasons of a Man's Life* at around the same time, also liked the idea of four stages, but with rather duller titles, and he also stopped at the fifties. A pity, too, that it was only men whom he interviewed, but this was the Seventies. Sheehy subtitled her book 'The Predictable Crises of Adult Life' and predictable the stages have always seemed, even if their details varied over the centuries.

Last month, however, I met, on separate days and by chance, a range of people who reminded me that there is more freedom to change the sequence than we might have thought. There was Peggy, a happy grandmother of 38, having had her first baby at 18, who then had her first baby at 19. 'I'm a daytime mother,' Peggy said, 'so that Katie [her daughter] can have the chances which I never had. She's studying design, you see, at the local college and Reg, her man, is off driving his lorry.'

Rebecca is two years older than Peggy and a lawyer. She is four months pregnant with her first baby, having decided that she could leave the start of a family no later than 40. She will find it hard, she said, to leave her stimulating job but she had decided that a baby should be a full-time job and she was looking forward to it. Robert, her husband, is none too sure about babies and the changes they will bring to their rather comfortable life style, but he does not intend to alter his pattern of life – he is an international banker, jetting round the world half the time. 'We shall see!' said Rebecca.

Joshua, another acquaintance, is different. He has 're-tired' at 37 in order to bring up his son, having been separated and then divorced from his wife. He is living in the West Country in a farm cottage, surviving on the rent of what used to be his London apartment. 'It is so much more fulfilling than all those meetings. I am really enjoying being a father and I suspect that I'm a much nicer person now than I used to be. I have taken up metal-working and will do it as a business once Harry is old enough to go to school.'

'Who said anything about retiring?' growled Lord King, aged 75, when he handed over his chairmanship of British Airways to become the part-time president. He might only come in two or three days a week but he would need the new space to give time to all his other activities.

Work's Variety

We could all add anecdotes of our own to this list. They are examples of the new flexibility which we have – to be a grandmother for the first time at 40, or a mother for the first time; to 'retire' at 39 or 75; to be a full-time parent or not. In fact, however, all that these people are doing is to define their work in different ways. Work is useful activity, and it comes in four varieties. There is, therefore, *paid work* in its various forms, be it for wages or for fees, depending on whether you sell your time or your produce, but there is also the *gift work* we do for free, for the community, for charities, sports clubs or political parties.

Then there is *home work*, not the preparation for school but the maintenance of the home and the care of the people in it, one's children, maybe, or ageing parents, or sometimes both. The Legal and General Insurance Company in Britain regularly estimates the replacement cost of a spouse as household-manager, for insurance purposes. In 1993 the figure was £18,000, well above average earnings that year in Britain. This, as women have always known, is real work. Fourthly, there is *study work*. In the knowledge age, the acquisition and development of intelligence, the new form of property, is an essential investment. It is also hard work. It should not be dismissed as a leisure activity or as something which is a tedious necessity at the start of life but unnecessary thereafter.

These extra categories of work are important. If, for instance, gift work and study work were accepted, officially, as 'work' in Britain, we would not have rules which prevent the unemployed studying full-time or working for more than 16 hours a week for a charity, on the theory that they are not then 'available for work'. They would *be* working. A balanced life is a blend of all these types of work. A full 'portfolio' has some of each most of the time. It will, however, be a changing mix as we move through life. We have more freedom than we think to change the mix.

Like the Hindu scriptures, I would settle for three active ages in life, with a fourth which is really the readiness to die, which we must not gloss over. The descriptions of student, householder and retirement are not, however, the right ones for the world we live in. Life is longer now, more things are possible. Women can bear children, safely, into their fifties. Fathers have it easier; they can and do start a new family in their sixties. We are likely to be as healthy and as fit at 70 as our parents were at 50. 'Retirement', therefore, can be a movable event if it literally means withdrawing from active work of any sort. Householding (and housekeeping) has still to be done but need not describe the whole of the middle of life for anyone. Over 80 per cent of women do some paid work these days, even though they still also bear the brunt of the home maintenance and almost all of the caring, the 'home work'.

The Four Ages are:

1 The First Age, the time of preparation for life and work, which includes schooling, further education and qualifications, guided work experience, and, I believe, the chance to explore the world beyond the home environment. The French word – *formation* – describes this period rightly. It is the age of 'forming' oneself, something which is more, much more, than formal education.

2 The Second Age, the time of main endeavour, either in paid work or in parenting and other forms of home work.
3 The Third Age, the time for a second life, which could be a continuation of the second but might, more interestingly, be something very different. To do nothing is no longer a realistic option.
4 The Fourth Age, the age of dependency.

Each age will, very roughly, last for 25 years, although I suspect that we shall see the Third Age stretching out to 30 years or more for many people. This age will only end when the Fourth Age begins, when we enter that anteroom to death. Naturally, the longer we can postpone the Fourth Age the better.

The First Two Ages

The success of each age depends very much on what went before. A successful First Age makes a Second Age much more likely to be successful in its turn, and that, again, helps to increase the options for the Third Age. It is crucial, therefore, that we see life as a whole if we are to use the opportunities of the three ages to build a cumulative balance to our lives. Most of us, we must remember, will, if we survive the dangers of the road and of drugs in our First Age, live on into our seventies or even eighties. The most difficult thing to do, however, is to think beyond our own experience, to conceive that there is a good life after 50, or sex after 40 when we are only 20.

The elders in society need to make it easier for younger members to see life as a whole and to prevent them, as far as possible, from mortgaging their futures too early. The 40 men whom Levinson interviewed for his book reckoned that life got serious at around 28, with a novice, or apprenticeship phase for the four or five years before. We may mature, physically, a bit younger now, but I doubt that many want to get serious about life and mortgages any earlier than they did.

For that reason we need to accept as a society that the First Age lasts for 25 years. It is already that length for the professional classes, for those symbolic analysts, the high-skilled knowledge-workers who make up 20 per cent of the population with 60 per cent of the income. They get their degrees and follow that with a professional qualification and/or a period of tutored apprenticeship. In Germany the process can continue until the age of 27, or until 31 if you want a doctorate. If we are going to spread the new bases of wealth, knowledge and intelligence more broadly, then everyone will need this extended 'formation' period. To shorten it for some would be to ration the new property.

It is odd, I often think, that we, the symbolic analysts, lament the 'idle scroungers' who are not actively seeking jobs at 18, but worry if we catch our own children wanting to dive into serious work at that age. 'But your degree!' we cry. 'And don't you want to see more of the world before you settle down?' If we want to see more of that potential intelligence wealth created we should start by applying the same standards to all and finding more ways to fill the First Age of preparation usefully for all. It would be a better investment for society to spend money early on our young of all classes. They might, then, be better able to fend for themselves in the Second and Third Ages. As it is, we skimp on the First Age and, as a consequence, spend more supporting many of them in the longer and more expensive ages which follow.

The Second Age is the age of serious endeavour. For many that will mean a job, for others parenting, for most a mix of the two types of work. It is in this age that the re-thinking of time becomes so crucially important. It is easy to get trapped in other people's time cages, be it the employer, the school, the shopping-routine or the parents' roster. This, too, is the age which has been sharply reduced for many with the advent of the 'compressed career', making the cage even tighter.

No longer can the average employee, as we have seen, look forward to 45 or 50 years of continuous work, the 100,000 hours. For those in the core those 100,000 hours may be compressed into 30 years, if they are lucky and can stand the pace. Many will not want to, some will not be up to it, some will burn out. Even if the organisations rethink time creatively so as to give people more control of their time, some may still burn candles at both ends for the corporation because the pressures to perform will be high. 'There is no way that I could do this job,' said one senior woman, 'if I wasn't prepared to give it all my time.'

This Second Age is an uneven balance for most of us. There is too much home work for some, too much time given to paid work for others, too little study work or gift work, too little or too much time without work of any type. It is easy to say that we ought to stand back, and we should look at how we live and try to rebalance our time and our work. It is less easy to do it. We are caught in the cages. For some, the cage is the workplace, for others the home. Necessity doth make prisoners of us all. If we feel like that we have to remember that it is not a life sentence.

The Third Age

The big change in recent times has been the gradual emergence of the Third Age, which will for many be the longest phase in their life but, paradoxically, the one for which they are probably least prepared. Because it did not happen to the generations ahead of us we did not expect it to happen to us, although, as *The Paradox of Age* noted, we know that every generation has a different life course.

The Third Age is not a synonym for retirement. For one

thing it starts too early for that to make sense. Some will extend their Second Age until their sixties; some, like Lord King, until their seventies. They will be the exceptions. Some women will find that the empty nest never empties, as ageing parents move in to replace the children. Most will find that, with the new compressed careers, the main work of their Second Age comes to an end in their fifties, or even earlier. No one seems to know what happens to the boys and girls of the dealer rooms who peak in their twenties. It is not ageist thinking to note that swimmers are past their best at 20 and tennis-players (even Jimmy Connors) at 35. Nor is it ageist to note that soldiers are less good at storming beaches in their forties, creative directors often less creative as they near their mid-life and that journalists must turn columnist or leave in their forties. They should not despair. The Third Age, whenever it starts, is the opportunity to change their blend of work, not to stop activity altogether. Some women, for instance, may want to increase the proportion of paid work in their mix, while their partners want, or are forced to decrease it.

Some paid work will be essential for nearly everyone in this Third Age. Necessity will make portfolio workers of us all in the end. Neither society, nor organisations, nor individuals have prepared properly for the support of their people for the 25 years or more of this Third Age. To the extent that this Third Age existed in previous generations, it was much shorter, at least for men. The Geneva Association, a research foundation of the insurance industry, speaks of the four pillars which are necessary for a financially comfortable Third Age. These are a state pension, a personal or occupational pension, personal savings or inheritance and some paid work, the fourth pillar. The state pension is going to get less and later everywhere. In Britain it will be down to eight per cent of average earnings by 2030 if it continues to be indexed to prices not earnings. In any case, 31 per cent of men retiring in Britain in 1991 did not qualify for a full pension because they had not contributed regularly over 44 years. Very few will meet that requirement in future.

The idea that one contributes to one's state pension during one's working-life, contributions which are then salted away for us until we need them, is, anyway, a form of deceit, in Britain and elsewhere. Our pension contributions are just another tax which goes into the general pool. It makes sense when you think about it. There is little point in the government putting our savings into a government fund which would then be mainly invested in government bonds, i.e. lent back to the government again. The pension is, therefore, an undertaking by those behind us to pay for our old age, the generational contract. That was fair enough when there were six of us working to every one of them, when they did not live too long or get too expensively ill, and when we were demonstrably better off than they had been. Those conditions no longer apply. By the year 2020 in most countries there will be only three workers to pay for every person over 65, and only two people in the Second Age for every one in the Third. There is no way that successive generations are going to agree to pay huge taxes to keep their elders as comfortable as they have been.

Few have noticed what a number of recent studies have confirmed, that for the first time in history, the elderly in North America, Europe and Australasia are, on average, receiving more income, spending more and saving more than people in their twenties and thirties, when responsibilities for dependent children are taken into account. This has come about despite the rise in the number of young women in employment, despite earlier departure from the workforce and despite the fall in the number of dependent children. This disparity will only increase. The new Third Agers are not the old of the welfare generation; many of them carry with them substantial assets from their Second

Age. In New Zealand the signs are there already. The median single-income two-child family has seen a 20 per cent fall in real purchasing power in the past 20 years, whereas the elderly have experienced a 100 per cent rise.

Given these sort of statistics, there is little doubt that the generational contract will gradually be renegotiated. Pensionable ages will be raised, as is already happening in Italy, America and probably in Britain (for women). It will go farther. Society will concentrate its money on the old old, not the young old. Those in the Third Age will, increasingly, be expected to make more provision for themselves. The fourth pillar of paid work will, then, be a financial necessity unless they have taken great care to build up their savings or private pensions in the Second Age. Those in the Second Age should take careful note.

We should not find the prospect of *some* paid work too daunting. Most people in the Third Age would like at least half a week of paid work and the new flexibilities of the workplace make that very possible. At the professional technical skill level, age is an irrelevance when you are self-employed and outside the organisation. Do you need to know how old your lawyer or your electrician is provided he or she can deliver what you need? At the semi-skilled level, age is sometimes regarded as a positive advantage. Supermarkets regularly report that part-time older employees are more reliable, less disruptive or ambitious and more friendly to the customers than the youngsters whom they can hire at equivalent pay.

The Third Age is, however, one's best chance to experiment with a different blend of the four types of work. It can go either way. Some, particularly women, may want the chance to increase the amount of paid work in the blend, if they have been largely restricted to home work in the Second Age. This would probably mean expanding their existing portfolio because they may find it difficult to enter the core of an organisation at an age when others are leaving. On the other hand, as the baby boom of the 1960s works its way through life, organisations are going to be less youth-conscious and more talent-conscious. Provided that the stock of intelligence, which the new recruit brings with her, is up-to-date and relevant, age may be less important than the new intellectual property. Once again, the use one makes of the Third Age is very dependent on preparation in the Second.

I like the story which my one-time publisher, Robin Waterfield, tells of Gerard Groote who was a successful professional cleric back in the troublesome times of the fourteenth century in Europe. He became the founder of a religious movement, called the Brothers and Sisters of the Common Life, which was strongly opposed to the hierarchical Church of that time. What interested me, however, was the way he decided to change his comfortable life and make his small mark on history. A stranger approached him one day, out of the blue, and said, 'Why are you standing here, intent on empty things? You ought to be another man.' The Third Age is, I feel, our opportunity to be another person. Not everyone has, as yet, got that chance. With four pillars of finance the Third Age is more than comfortable. With only one pillar, and that the diminishing state pension, there is little scope to be anyone else. I shall argue, in another chapter, that social justice requires that we make that choice more readily available.

Life is long, for most of us. In each stage of that life there are difficult choices to make. In the Second Age, particularly, it is easy to lose oneself in busyness or in emptiness. I have met as many empty raincoats rushing hither and yon as I have seen huddled lonely in a corner. Properly used, the First Age of 'formation' should be a time to grow an identity before the consuming busyness. Properly balanced, the Second Age is the time of one's major contribution, to work or home or community. The Third Age is the opportunity to be someone different, if we want to be; or to go on doing what we used to, only slower. If it is true, and it is unverifiable, that by the time we die most of us have only discovered one-quarter of what we are capable of doing and being, this is the age to find the missing three-quarters.

Because their numbers will be so big – one-third of the population in most countries – and because they will be spending and saving more than those a generation behind them, and because they will bring with them all the expertise and contacts of their Second Age, the members of this Third Age will be a powerful influence in our societies. Their values, their money, and their votes will count. It is not a homogeneous group, nor is it or will it be organised, but if their spending is on time not things, time to travel, time to study, time to eat and time to watch, they will alter the pattern of work for the rest of us. It will be even more of a service and a knowledge society. If they also prove, in this stage of their lives, to be more concerned that things should be better not bigger, they will affect our priorities. The environment might get more practical attention, and more money; town-planning might be more people-friendly than car- or store-friendly; consensus might become preferred to confrontation. On the other hand, given their influence, and their votes, the Third Age population could pull down the shutters on change, huddled in their ghettoes of rich and poor, and let the world go hang.

The Fourth Age

The Fourth Age presents the biggest challenge to the idea of a balance in our lives. The questions are only beginning to be explored. When is a life not worth living, and by whom and how is that decided? The Dutch are the first to have licensed euthanasia under certain specified conditions. Others will follow. Some worry that we shall always get the balance wrong on that, but doing nothing may also get the balance wrong. Maybe the old Hindu tradition that one should go into the forest to die when one's life was over was a good one. I have always thought that I would like to attend my own memorial service. That way I could, before I left for the forest.

The more urgent question is the cost of one's final days, and sometimes years. More health-care money is spent on the last year of our lives, on average, than on all the rest of life. Can that be a sensible balance? If it is public money that is so used, then it is at the expense of better care for generations behind. The trade-off or compromise is hard to make, but, again, no decision is still a decision. It is, I suspect, impossible to find that compromise without a clearer consensus on the meaning and purpose of life, and death.

Splicing It Differently

Most people will progress through the stages in due order, using, if they can, the Third Age to fill out those parts of their lives which were left untouched in the first two, and postponing as long as possible the arrival of the Fourth Age. It could, however, be different. Many women would like to have the opportunity of a Second Age of paid work and a serious career in their late forties and after, the time when most men will, in future, be entering their Third Age. With the new flexibility of working time, there is more and more possibility of this happening. Others would like to postpone the Second Age more or less indefinitely. Attracted by the possibilities of the Third Age, they would like to bypass the Second Age and enter the Third as soon as they have finished their schooling. Provided that they do not expect society to, as it were, pay their pension in advance, but are able to support themselves on the fourth pillar, there is no reason why they should not do this. We do not all have to be successful careerists in our middle time.

The contradictions and paradoxes of life cannot be removed. We will seldom be able to get all that we want out of life at the same time, nor be able to give to it all that we might like, at one stage. The solution may be a Chinese contract, a trade-off between what you dream of and what is practical, but it may be a third angle, taking and giving different things to life at different stages.

One of the more moving moments of my life was watching a degree ceremony at Britain's Open University. It was held, appropriately I thought, in a cathedral, because each graduate was there because they had made an effort towards some form of personal renewal. What was moving was the huge variety of the graduating class. Grannies were there, and great-grandfathers, photographed by their progeny in their caps and gowns with their degree certificates, instead of the other way round. There were people in wheelchairs and others with guide dogs. Age was no barrier in that place, nor class, nor creed, nor colour, nor previous success in anything, for it is a truly 'open' university. It was, for me that day, a splendid example of the infinite possibilities of life. That First Age of formation and learning can come again at any age.

Changing stages is, however, a version of the second curve. There is always a downward beginning before the upward swing. I have, myself, changed careers three times, from oil executive to academic, from tenured academic to life on a priest's salary, and finally, or perhaps not finally, to working as an independent writer. In every case my finances took a dip. Each time I was again the new boy in an alien world and had to earn my credibility anew. Each time, however, the difficulties decreased with time and a whole new life emerged. Others have been even more courageous, forsaking the office for a circumnavigation of the globe, starting farming in mid-life, reversing roles in the family in mid-stream, moving from priest to advertising executive, from nursing to founding a software consulting business, or from chief executive of one such to a full-time artist.

Life is full of possibilities for a second curve. In searching for it, we will need some sort of core to our personal doughnuts, some basic wherewithal, but it can be smaller than we think. Some Chinese contracts with those around us will often make it possible for everyone involved to mix it differently. The first three ages of life can be lived in any sort of order, and the new chaotic state of our world makes it unusually possible to be different from the norm.

A State of Justice

The individualised working lives foreseen in the last chapters could fill out the raincoats for some but could also leave many of them emptier than before, bereft even of a place of work to go to, no matter how boring or anonymous, and lacking the means and the know-how to be an independent individual. The spectre of a divided society looms over us, no matter how federal we are or how well-meaning our businesses. We have to tackle the paradox of justice and, in particular, the paradox of intelligence. If we don't, we may well bring the whole edifice down about our ears, because it is ultimately not tolerable for the many poor to live beside the fewer rich. It would, anyway, be crazy, as well as immoral, not to want to create a full property-owning democracy when intelligence is the property, because of the happy paradox that more intelligence for some does not mean less for anyone else.

These chapters are not intended to be a discourse on the nature of justice. Justice is, however, the bond of society. Justice allows us to dwell together in unity, building a beneficial compromise between the rights of the individual and our responsibilities to our fellow human beings, enabling us to love both ourselves and our neighbours. If we want to avoid that spectre of a divided and embattled society, we should do our best to create a state of justice in our land.

At its simplest, justice means fairness. Fairness means, for instance, that society should not deal with people arbitrarily but with 'due process' – this is the legal side of justice, which need not concern us here. Fairness also means that not everyone should get the same, because not everyone needs or deserves the same. In practice, anyway, a strict equality does not work. As Abraham Lincoln said, you don't make the poor rich by making the rich poor. Fairness could mean, however, either that we gave the brightest of our young the best of our education because they would make the most of it, or, conversely, that we gave the least talented the best because they needed it most. Fairness is always a complicated question.

Fairness, when it descends from lofty principles to hard decisions, always means a compromise, a blend of two 'oughts'. In the case of education, for instance, fairness means that everyone should, as far as is practically possible, have the same chance to be different. We should not tilt the scales against anyone from the start. We should also give people the chance of more than one start if they are slow off the mark. On the other hand, we should also encourage with our help those who make the most use of that early start. No one should want to cut back the education of doctors to create more schools for delinquents, on the grounds that the latter need it more. Justice always seeks to balance the needs of the individual with the needs of the wider community.

In the context of the issues discussed in this book, fairness means giving everyone a decent chance of a life on the second curve. In Britain, the Commission for Social Justice put it this way in their first report in 1993: 'we cannot help but regard a commitment to the extension of opportunities as a radical doctrine, and one that lies at the heart of social justice'. In a democracy where wealth derives from property, fairness, therefore, means giving everyone a chance to get some of that property, which, in the new millennium, means intelligence of one sort or another. Fairness therefore requires an **investment in intelligence**, an investment in the education of all people throughout their lives, accepting that some people will make more of that investment than others. Chapter 12 investigates what that might mean – the different types of intelligence, the forms which education might take, and the help which people might need to develop their skills and aptitudes as they go through life. In a state of justice, everyone has a right to some property. What use they make of it, however, is up to them. We must each take responsibility for our own doughnuts in life.

Fairness also suggests that there should be more chances to win in life than to lose. A third angle on this problem is to propose that there should be more than one measure of success. Where there is only one scale, there will always be winners and losers, and usually more losers than winners. In a contented society, with more winners than losers, there need to be multiple scales, a variety of ways to feel good and to count yourself successful. Society will then have more givers than takers, and a greater variety of life.

Chapter 13 looks at some of the alternatives to money as the measure of all things, and proposes a **new scoreboard**. What is counted is what counts. It is not enough, therefore, to say that a good life is more than jewels, or that the environment is important to us all. We have to make some sort of stab at measuring such good intentions or that is all that they will remain.

More measures will mean more compromises between the numbers on the different scales, in personal life as well as in business and in society as a whole. The principle of the doughnut, that there are the necessary things of life, the core, and the other things, which are the ones which make the difference, provides one path to that compromise. There is, however, no one answer for all. Justice requires that we eliminate the worst inequalities where we can. Justice does not require that all should be the same. That, in fact, would be unjust, a denial of our right to be different within limits.

12 The Intelligence Investment

When Intelligence is Property

In a property-owning democracy which claims to be fair to all its citizens, it is only right that everyone should have a share in that property and the wealth which it brings. When property meant land, social and political revolutions redistributed that land, most recently in bits of ex-colonial Africa. When property meant stocks and shares and the ownership of enterprises, governments went out of their way to encourage more of their citizens to catch the shareholding habit. Alternatively, governments sought to persuade them that nationalisation was one way to give all citizens a stake in the property of the nation.

Now that intelligence has replaced land as the source of wealth we have to take seriously that opening sentence of *A Nation at Risk*, the 1983 report on American education: 'All, regardless of race or class or economic status, are entitled to a fair chance and to the tools for developing their individual powers of mind and spirit to the utmost.' If we don't make this new property more widely available, if we don't invest in the intelligence of all our citizens, we shall have a divided society.

You can already see that divide deepening. Robert Reich has divided the modern American workforce into three categories: there are the routine operators, who are still needed to pack the airline meals, operate the tills and put the data on to the discs. They make up maybe one-quarter of the labour force, a proportion which is declining as their jobs get automated or are exported to lands with cheaper labour. Secondly, there are the personal-service providers in restaurants, hospitals and security, 30 per cent and growing. Thirdly, there are those people whom Reich calls the symbolic analysts, those who deal with numbers and ideas, problems and words. They are the journalists, the financial analysts, the consultants, architects, lawyers, doctors, managers, all those whose intelligence is their source of power and influence. They now make up perhaps 20 per cent of all workers. Farmers, miners and government employees make up the rest. It is the symbolic analysts, the knowledge-workers, the professionals and the managers, who are the real beneficiaries of the information age because they own the new property.

Under present policies this 'fortunate fifth' is getting richer almost by the minute, while the others get poorer. Reich calculates that in 1989 this top fifth had a higher after-tax income than all the other four-fifths combined. In times gone by, the rich had a vested interest in supporting the poor – in the final analysis the poor were both their customers and their neighbours – but the new rich, the symbolic analysts, sell their stuff to each other, or to other firms, internationally. They do not venture downtown, use public transport, or send their children to the public schools. Why then, they say, should they pay more to support more of such things? They do not benefit themselves, even indirectly.

The conventional wisdom has been, both in the United States and Europe, that the private sector pays for the public sector. Help the private sector to get rich and the other sector will benefit. Selfishness makes sense. That was true when property was the old-fashioned sort – land, bricks and machinery. Wealth did trickle down. More of that sort of property needed more people to work it. Intelligence-as-property changes that beneficial sequence. The causal chain is reversed; a rich private sector no longer results in a richer public sector, it goes the other way round. Without investment in the public sector, in housing, in telecommunications and transport, and, most of all, in education, the number of those symbolic analysts cannot increase significantly; the stock of useful intelligence will remain confined to one-fifth of the population. The rest will be progressively cut off from the world of property in the new sense, increasingly poor and effectively disenfranchised.

The Nature of Intelligence

If intelligence is the new basis of property and wealth, it is odd that we don't always seem more eager to grab more of it for ourselves. In Britain nearly three out of 10 youngsters leave full-time school as soon as they can, at 16, without any qualification and often without any educational certificate in any subject. We know, by contrast, that in Germany, Japan, the Netherlands, France and America, 90 per cent stay in school or formal training until at least 18. In America, however, it may not do them all that much good. As the Education Committee of Congress discovered, fewer than four in 10 young adults can summarise in writing the main argument from a news column. Only 25 out of 100 young adults can use a bus schedule to work out how to get from here to there at a particular time. Only 10 per cent can select the least costly product from a list of grocery items on the basis of unit-pricing information. Something isn't working as it should. Either the young are short-sighted and stupid or, just possibly, they are right – they don't feel that they are learning what they should while they are at school, not in Britain or America, at any rate. It is not, they may instinctively feel, the right sort of intelligence in which to invest.

Consider, on the other hand, this question from an entrance paper at Tokyo University.

> Given a regular pyramid V with a square base, there is a ball with its centre on the bottom of the pyramid and tangent to all edges. If each length of the pyramid base is of length *a*, find the following quantities: (1) the height of V; (2) the volume of the portion common to the ball and pyramid.

How many of our students applying to study maths at university, one wonders, would be happy to tackle this problem on their way in? The snag is – this was not a question in the maths entrance exam, it was in the paper for humanities students! The academic standards are high in Japan. None the less, when these sophisticated learners start work, they have to start learning all over again. The businesses of Japan see the universities as a recruiting-ground, not an education. As they used to say of Oxford, it only needs to run a recruitment office and a placement office; what happens in-between is irrelevant.

The Japanese themselves worry that their educational system is no longer preparing people adequately for a complex and shifting world. Other countries are also puzzled as to how best to deliver this new form of property. Intelligence may be the source of wealth, power and freedom, but, inconveniently, real intelligence is not a substance, it cannot be pre-packaged, sorted and delivered as if it were a consumer product. Some elements of it can, it is true; intelligence defined as information can be treated in just that way, it can be pre-packaged, disseminated, stored and retrieved; it can be mass-produced, made consumer-friendly, distributed in multi-media and tested for reception. It is very tempting to think that when that form of intelligence has been dispersed, the job has been done. But to know all is not to be able to do all.

I have long admired Howard Gardner's concept of multiple intelligences, as described in his book *Frames of Mind*. In that book he lists seven intelligences and describes how they can be measured. He arrived at his theory by watching brain-damaged patients. Some had a normal intelligence but could not remember their personal history, or recognise faces, even their own. Others could do everything except count. The important conclusion is that none of the intelligences is necessarily connected with any other. You can be as bright as a button in one and a dunce in another. You can shine in five or only in two. My own list has nine different forms of intelligence:

Factual intelligence: the sort of walking encyclopedia who wins the Mastermind competitions in Britain, who knows the answer to every question in Trivial Pursuit and can give an impromptu lecture on the state of the Romanian economy over dinner. We are envious but often bored.

Analytical intelligence: the person who loves intellectual problems, crosswords and puzzles. Such people delight in reducing complex data to more simple formulations. Strategic consultants, scientists and academics are strong in this type of intelligence. When this intelligence is combined with factual intelligence, examinations come easy. When we describe someone as an intellectual it is often this combination which we have in mind.

Linguistic intelligence: the one who speaks seven languages and can pick up another within a month. I envy such people, since I don't have this facility myself, but we have to remember that it is not necessarily connected with the first two intelligences.

Spatial intelligence: the intelligence which sees patterns in things. Artists have it, as do mathematicians and systems-designers. Entrepreneurs have it in dollops, but without necessarily having the other intelligences, which explains why many an entrepreneur failed at school and would never go near a business school.

Musical intelligence: the sort that gave Mozart his genius, but which also drives pop stars and their bands, many of whom would never have a chance of going to college, because their scores on the first two intelligences would have been too low.

Practical intelligence: the intelligence which allows young kids to take a motor-bike apart and put it together again, although they might not be able to explain how in words. Many 'intellectuals', intelligent in the first two sense of the word, are notoriously impractical and unworldly. 'Am in Crewe', Chesterton cabled his wife. 'Where should I be?'

Physical intelligence: the intelligence, or talent, which we can see in sport stars, which enables some to hit balls much better than others, to ski better, dance better and generally co-ordinate mind and muscle in ways that defeat me.

Intuitive intelligence: the gift which some have of seeing things which others can't, even if they cannot explain why or wherefore. It is said that women have this intelligence to a greater degree than men, which may be why men often disparage it.

Interpersonal intelligence: the wit and the ability to get things done with and through other people. Notoriously, this intelligence often does not go with analytical or information intelligence. 'Too clever by half' – the jibe aimed at the Conservative politician Iain Macleod in years gone by, to explain why he would never be the great leader he could have been, applies to others as well. Without this form of intelligence, great minds can be wasted.

My list is based on observation. There may be more types of intelligence than nine. The important point is that intelligence has many faces, all of them useful, all of them potential property in this new world of intelligence. We will not all be symbolic analysts in the future, but we will all have to create and manage our own work doughnuts. To do that we need to have a clear idea of our best intelligences, and have learnt to make the most of them. It may be more an article of faith than a researchable fact, but we should make the starting assumption, in a just society, that *everyone* is intelligent in at least one of the nine ways. It should, then, be the first duty of any school to discover one's intelligence(s) and develop them. 'Know Yourself', said Juvenal, were words given from the gods, and inscribed on the ancient temple of Delphi. An impossible precept, grumbled Carlyle, it should be replaced by the more nearly possible 'know what you can work at'.

The Three 'C's

Discovering your intelligences is one thing, applying them is another. We need to be able to recognise and identify problems and opportunities. We need to be able to organise ourselves and other people to do something about them, and we need to be able to sit back and reflect on what has happened in order that we can do it all better the next time round. It is the cycle of discovery at work.

The skills involved are conceptualising, co-ordinating and consolidating – the three 'c's. They are the 'verbs' of education as opposed to the 'nouns', the 'doing' words not the facts. We don't learn to use these verbs by sitting in rows in a classroom, but by practice. Without them we may be a potential Nobel prize-winner or a star athlete but no one, least of all ourselves, will ever find out. These three 'c's should be the core of any educational doughnut. Unfortunately, they seldom are; they are regarded instead as add-on extras, optional skills for what space is left in the doughnut. That is why the businesses in Japan have to re-educate their clever new recruits as soon as they arrive. That is why some kids may be right to leave school early – they will learn the three 'c's more quickly on the streets.

I asked a professor of English at Cambridge University what they did there to educate his students for the demanding and prestigious jobs which most of this talented group would surely move on to in life. That, he said, was no business of his. 'They come here to read English, and that is exactly what they do.' Tony Benn once listed his education in *Who's Who* as taking place 'in the intervals between terms at Westminster School and Oxford'. He may have been right. The children of the symbolic analysts learn the three 'c's as they grow up, mentored and coached by their elders as effectively as the new recruits are in the Japanese firms. Their parentage thus accentuates their advantages.

A just, and sensible, society will do something about that accumulating difference between the children of the successful and the others. Since 80 per cent of the young do not have symbolic analysts as parents we have no choice but to use their early schooling as a substitute. That means intensive care and attention in the years from four to 10, when the 'c' skills are beginning to be formed. At present, in Britain, there are 25 children to every teacher at this level, but only 10 students per teacher at the undergraduate level. We ought to reverse the ratios.

If they were properly educated at the start, students ought to be able to take responsibility for more of their own learning at the university level, something which they are unfairly expected to do at the primary stage. When it is argued that there is no evidence to show that class size at the primary level affects learning, I have to point out that what is being measured in that research is the retention of information or the acquisition of repetitive skills, the 'nouns'. The 'verbs', as we know from trying to develop them in adult life in organisations, need mentoring, small-group experience and real-life problem-solving. You have to live them to learn them.

With 10 small children to a teacher it is possible to approximate the kind of real-time, real-life learning that the children of the symbolic analysts pick up. With 10 children it is possible to move between the classroom and life outside in a way which is not logistically possible with 25. It can be any 10 children. There is a lot of piecemeal evidence to suggest that you do not have to be the child of a symbolic analyst to learn these things. Most people can do it if they start young enough.

In a famous programme in America Jaime Escalante got low-income Hispanic students through the Advanced Placement Examination in calculus, one of those conceptual subjects normally restricted to the so-called 'brightest' in the class. If he could do that with one of the toughest of noun skills there is no saying what he might have done with the verb skills. What small groups and close mentoring and learning from life can do, and that which large classes can rarely do, is to give a child self-confidence.

Portfolios at School

We could go further. Instead of requiring the student to reach certain standards before she or he gets their leaving certificate from school, we could require *the school* to ensure that the student has reached those standards before they let them go. School should be a place for compiling a portfolio of competences. Those competences need not, and should not, be age-linked, with levels or tests or examinations to be passed at particular ages, because people learn these verb skills at very different paces. Like music tests or driving-tests they should be taken when one is ready for them and likely to pass. If every 16-year-old takes the same examination at the same time, and if that examination is graded, half of the population will, inevitably and logically, do better than the other half. The net effect is to persuade half of the population that they are failures, however often you tell them that they have passed.

Wherever we need tests we must make a distinction between age and competence, and allow retakes of everything. In the end, almost everyone everywhere passes their driving-test. My daughter passed hers at 18, my son at 24, because he was not in such a hurry to learn to drive. Neither of them, now, a year later, thinks that they are a better or worse driver than the other. If everyone took the same driving test at 18, and only the top half were considered competent to drive, we should have fewer, better drivers and safer roads. We should also have a lot of very deprived and discontented people, including many who might well have developed into very competent drivers a few years later.

If the age-bonding of our schools went on throughout life, we should be very resentful. If only 25-year-olds could take the accountancy examinations or only 39-year-olds apply for full professorships, there would be an outcry. It can only be for reasons of administrative convenience that schools remain the most ageist of all our institutions. The effect is to make half of our young feel that they are failures.

Students should each be required to compile a bulging portfolio of certificates of competence or achievement. Apart from certificates of competence in the traditional subjects, I see no reason why driving, swimming, first-aid, word-processing, cooking, tax law, telephone and presentation skills, and any other practical life-skills should not be certificated and collected during this period of life. These are certificates of competence. They can be formally tested, as with music or driving, or they can be examined on the evidence of their work, as with artists' portfolios.

This form of portfolio collection should and does go on throughout life, but the habit needs to be acquired young. It will be an essential part of the portfolio life which we are all going to experience at one stage or another. Even inside the organisation, as I have argued, portfolios will be the way people develop their careers with the organisation encouraging them to add to their credentials at every level, sometimes by new accredited experience, sometimes by certificated tests or courses. The Records of Achievement which are becoming increasingly common in British schools are a step in the direction of portfolio collections. To be effective they need to become part of a nationally accepted scheme of educational portfolios, not the icing on the cake that they are today, the sop for those who cannot excel at the traditional examinations. When we make all examinations age-independent and when we happily include certificates relating to all the intelligences, we shall begin to see a proper balance in our education.

There is, however, no need for all this portfolio collection to take place *in* school although it should happen while *at* school. Nobody would seriously expect the school to teach driving or, perhaps, word-processing. The school can and should be the organising hub for all the extra-curricular activity. Come to that, there is no reason why subjects like languages or domestic science should not be taught by specialist agencies, under the general supervision of the school.

If the education service is unwilling to see its role so enlarged we should develop a separate Youth Service which took over where the formal schooling left off, handling all the sports, work experience and community activities as well as the more practical aspects of the portfolio. School proper might end, as on the Continent, at 2 p.m. when the Youth Service would take over, staffed by some full-time professionals but with the help of many part-timers, voluntary workers or parents, portfolio people themselves.

It could go further still. Technology, and the possibilities of multi-media, will make independent learners of us all. There is no reason why some may not choose to learn for themselves by themselves in some topics, presenting themselves for examination when they are ready, rather as one already does for a driving-test. The function of the school, or the Youth Service, would then be to act as a tracking-station to make sure that no one was falling through the net, or losing the benefits and lessons of the three 'c's. The school would then be the core of an educational doughnut organisation. Some teachers would be core staff, well paid for long hours and flexibility. Others would be specialists, working outside the core and selling their expertise to a range of schools or institutions, portfolio people themselves. Some would move between the two roles during their career.

The Double Bond

Portfolios and the doughnut school may not, however be enough. We learn about life from life and we learn about work through working, mixed with a judicious amount of coaching, teaching and reflection. The German model of a blend of workplace experience and formal instruction for all but the most academic at age 16, has been widely admired and is beginning to be imitated in many countries with subtle national variations. There is, however, always a danger that this is a recipe, in a changing world, for training people in jobs and skills which will soon be obsolete. Unfortunately, if you go along with curvilinear logic, all is ultimately obsolete.

This kind of learning must, therefore, be complemented by some of that 'verb' learning which might, in a sensible world, have been learnt earlier but probably won't have been. We should, therefore, present every young person on adulthood, at age 18, with a double bond. One part of that bond would guarantee to pay the fees, up to a defined level, with basic maintenance, for two years or the equivalent, of full-time study at any recognised learning institution. This part of the bond could be used at any stage of one's life. There would be no age barrier. It would be up to the individual to apply and up to the institution to accept or reject them. The state would guarantee payment but not admission. The assumption would be that sufficient institutions would create themselves to meet this demand once it was seen to be underwritten to this extent.

This part of the double bond would automatically be taken up by those who were going on to university from school. They would, therefore, get the first two years of their higher education free. If their course lasted longer than two years they would have to pay the cost of the extra time. At present, first-degree courses in Britain are three or four years long. If the proposal of a double bond were introduced there, we might expect to see the first degree compressed into two fuller years, to be followed by an optional two years of graduate study. These graduate years could be paid for by a graduate tax on further earnings. Those who earn less would then, automatically, have more years in which to pay it off. It is fair.

The other part of the double bond would be to guarantee to find a full-time job for any citizen who presented themselves, for two years in the local region, either in a voluntary organisation or in a government agency, at a level equivalent to the minimum wage, where such a thing exists. It would be a recognition that there are many things about life and work which you can only learn by working and living – the 'verb' skills in particular. The government would organise what would, in effect, be a brokerage or employment agency in each region. Since this would be labour in excess of the requirements of the labour market, the jobs would have to be located in the voluntary sector or in non-statutory government work where they did not substitute for longer-term jobs and workers. The two bonds could be used in conjunction, if the individual so wished and could so arrange.

This double bond would be a recognition of society's continued investment in every individual citizen when they reach adulthood, not just in those who see themselves on the academic track. In return, society would be entitled to refuse to support anyone until the two bonds had been used, accepting always that there will be exceptional circumstances in some individual cases. The double bond would be one way of providing the essential extra investment necessary to launch the individual into the 'intelligence society' with the skills, both intellectual and practical, which he or she will need to survive.

Although only a minority of people would, in fact, cash in their bonds, although the bonds would effectively replace many welfare payments, and although they would allow work to be done in the community which might not otherwise be affordable, the scheme would, potentially, cost a lot of money. It would have to be seen as an investment in the long-term future. Its pay-off would come in the reduced need to provide for these people in later life, if they were, as a result of the bonds, more able to look after themselves. It would come, indirectly, in the lower costs of policing and repairing a more contented and just society. If Singapore can think it right to invest 25 per cent of her GDP in education, training and development the rest of us should be able to do at least as well. It is naïve to think that learning for life can finish at 16 or even 18, yet that is the implicit message which we are giving today to many of our young.

Learning, like life, goes on for ever. It would be reasonable to expect that the workplace would see the good sense of investing in intelligence, at least for its core workforce. We would be often disappointed. Too many organisations do not think far enough ahead to wait for the pay-off from the investment. Others hope to cash in on other people's investment and entice their educated and trained staff to join them. Some rely on the individuals to invest in themselves. Some compromise is needed if learning is to become the fashion throughout life.

One way to encourage the fashion would be to set a legal benchmark for organisations, requiring them to spend a set percentage of their pay-roll on education and training, a figure to be reported in their audited accounts. Any organisation falling below the standard would forfeit the difference to a central training fund. The French require 1.2 per cent of the pay-roll to be spent in this way. Most firms exceed it. With the minimum of bureaucracy a minimum level is thus established, but to set the level at 0.5 per cent, as the British Labour Party recently proposed, is to underestimate the investment required in the age of intelligence. My university requires that an average of one day a week be spent on research, keeping ahead of my subject. That is 20 per cent of my time. One half of that – 10 per cent – might be a minimum standard for anyone in the years ahead. Five days a year, the norm for good employers, leaves a large gap to be filled.

Some of this money and time could be seen as the entitlement of the individuals, to invest in their own development as they thought fit. The organisation does not necessarily or always know what is best for one. If an individual is entitled to annual holidays, sick leave and maternity leave, it would seem only sensible to extend the entitlement to intelligence. The slogan of some American corporations – 'individual initiative and corporate support' – has the right ring to it, but is usually interpreted only to mean attendance at some selected courses. It needs wider application – a guaranteed sum of money per annum, accumulated for up to seven years if need be.

It is, however, the outsiders who are most likely to miss out on any continuing investment in their developing intelligence. Most organisations will leave it up to the individual, and most of those will be too poor, too busy or too short-sighted to do it for themselves. Here lies the biggest danger of the intelligence age, a diminishing competence at the bottom end of the labour market. Outsiders need help.

The New Agents

All independents need an agent. A good half of us will be independent at any one time in the future, and all of us will be independent at one stage or another. Independents are never unemployed, officially – they just have no work. Resting, they call it in the theatrical world. If the unemployment figures eventually decline in Europe it will be, in part, because many people will perforce have gone independent, have got a small portfolio and so taken themselves off the unemployment registers. To make the portfolio bigger and better they will need an agent. While some independents will have bulging portfolios and full diaries, many will be among the most vulnerable of our societies, unprotected, unwanted, deteriorating assets. That will be in no one's interest.

Actors and models have agents, writers have agents as do golfers, tennis-players and boxers. It is hard enough to market and price yourself when you are a star; it is impossible when no one knows you and when you are unsure of what you can offer. A good agent will not only find buyers for your talents and negotiate the deal, he or she will be a coach or mentor, helping you to review your experience and guiding you to appropriate educational opportunities. Good agents will prod your creativity by floating ideas in front you – 'Have you ever considered . . .?' or 'Would this sort of thing interest you . . . ?' They will suggest what you need to do or where you need to go to improve your skills or to enlarge your experience.

They will also, if they are any good, help to organise your life so that there is some order in the necessary chaos of the independent's schedule. This is not altruism. It is in their interest to increase the value of the asset they are managing. It is, for the independent, a great comfort to know that there is someone whose interests entirely coincide with theirs, because it can be a lonely world outside the organisation. There is a growing market for agents of portfolio workers. The executive-leasing agencies have been quick to move in to the upper end of that market. They offer to provide executives to organisations to cover short-term skill gaps or project-managers. They are, in effect, agents for portfolio executives.

The need is critical for those lower down the skill range. This would have been a natural opportunity for the old trades unions, whose membership and influence has inevitably waned as the minimalist organisation has waxed. The unions, however, have been noticeably reluctant to recognise this new market. We must therefore look to new intermediaries. It would be pleasing if the employment agencies were to be more than brokers, and were to live up to their name and act as agents, not of the hiring organisation, but of the individual. If they were far-sighted enough, it would pay them to spend money upgrading the skills and the knowledge of those on their books in order to increase the rates which they could then demand from the hiring organisation. Some are already beginning to offer training opportunities; more must follow suit.

Portfolio workers need more than agents, they need somewhere where they belong as of right. Learning is alienating if you do it all by yourself. Teleworking is fine in technological theory but lonely in reality. That asset which is yourself can atrophy in isolation. We independents need somewhere other than the home, somewhere where there are colleagues not clients, somewhere where we can find the companionship and gossip of the old office or factory but without the boss. Somewhere where we can exchange experience and contacts. We need a club. I have argued, earlier, that the hub of the minimalist organisation will be a clubhouse for the members of the dispersed core. It should also be available to key portfolio workers to use when they need it. One piece of everyone's portfolio should, if possible, include the use of a club facility as part of the fee. For the many who cannot negotiate that privilege, I would like to see the employment agencies begin to offer a similar facility in exchange for an exclusive right to sell your skills.

As the portfolio market becomes more competitive, we may see these new intermediaries actually employing a reserve labour force which they then sell on, keeping the risks and rewards for themselves. The portfolio worker would then be trading some freedom for more security, the guarantee of training and traditional employee benefits such as holiday pay and sick leave. Some organisations, Hewlett Packard in France for one, IBM in London for another, do something of the sort for their newly redundant or retired workers, putting them on a retainer or a guaranteed-fee basis for a certain proportion of their time. They have created clubs which pay you to belong to them, but to keep your membership you have to keep your skills up to date, you have to keep on growing your asset.

Some portfolio workers form their own clubs or networks. Networks are useful, but if they reside mainly in your address files they lack the spontaneity of a club. An address book is not quite the same as a bar and a reading-room. Every network needs a club at its hub to add the human face to the electronic impulse. Clubs for the unemployed offer the right facilities but can, too often, be places of shared misery rather than shared learning. Only if their members start to think portfolio does the club take on a new life, looking, now, not for jobs but for customers.

The independent workers of our societies are among the most vulnerable and the least protected. Europe's Social Chapter is an attempt to remedy that, but will, one suspects, be more often honoured in the breach than in the observance. Britain's refusal to sign it may only be the more honourable face of non-compliance. In a competitive world, with a surplus of labour sloshing around, independents will need all the help that they can get. It is in all our interests to give it to them.

13 The New Scorecard

Unfortunately, Macnamara was right. He said, in what has come to be known as the Macnamara Fallacy:

> The first step is to measure whatever can be easily measured. This is OK as far as it goes. The second step is to disregard that which can't be easily measured or to give it an arbitrary quantitative value. This is artificial and misleading. The third step is to presume that what can't be measured easily really isn't important. This is blindness. The fourth step is to say that what can't be easily measured really doesn't exist. This is suicide.

What does not get counted does not count. Money is easily counted. Therefore, all too soon, money becomes the measure of all things. A just society needs a new scorecard.

The Distortions of Money

It is not always realised that the idea of national income-accounting, on a regular standardised basis, the GDP and the GNP numbers which we all now assume is what we mean by 'national income', is quite a new idea. In Britain it started in 1940 when the government, helped and advised by Keynes, needed to work out how much money they could raise in order to fight the War. Before that time there had only been occasional and non-standardised estimates. My first job with my oil company was in Singapore, where, in default of anyone else, they appointed me, trained in classical history and philosophy, to be their first 'regional economist'. I was asked to prepare a series of forecasts relating oil consumption to national income, drawing on ratios established elsewhere in the world. Unfortunately there were no national-income statistics for Singapore. This was in 1956, when it was still a British colony. I may have been the first to make a very rough and inadequate estimate of Singapore's GDP, an estimate which, I recall, involved guessing at the earnings of Singapore's colony of prostitutes.

Things are different now. Singapore is proud to boast of her per-capita income. League tables of national income proliferate. It is assumed that these equate with standards of living, but the statistics measure only the visible transfers of money. So, for instance, and most notoriously, they don't measure the unpaid work in the home. If, however, the wife dies and her husband hires someone to do the work which she did for nothing, the apparent prosperity of the country would rise by the £18,000 which the Legal and General Insurance Company say it would cost in the 1990s. Voluntary and charitable work – gift work – is not included because no money changes hands, nor the caring of the elderly if it is done for love or compassion in one's own home. Put your parents in a home for the elderly, however, and society, by these accounts, is immediately the richer.

More insidiously, if the cars and the highways are so bad that accidents proliferate, then hospital, car-repair and insurance bills increase, and so does the supposed wealth of the country as these transactions find their way into the national accounts. You can spend money polluting the clean air of the countryside with a factory, muck up its rivers and destroy the peace and stillness of the place, and it will all be counted as an increase in wealth because nothing is deducted for the damage. If the firm were fined, or charged, for what they had done, it would, apparently, make us even richer. We are encouraged to be a disposable society by the way we count. The more you throw things away and buy new things instead of having them repaired, the richer the society appears.

The distortions go on. Leisure, the precious commodity, only gets counted if you spend money on it. I have sometimes, half jokingly, suggested that the reason that the Germans are richer than the British is because the Germans tend to live in apartments while the British like their homes to have gardens. If you live in an apartment, every time you go out you will, normally, either spend money or make money. Meanwhile the Briton goes into the garden and watches the cucumber growing, or weeds the flower-bed. No money, so no wealth. Love is for free, so buy diamonds instead – it will make the country richer. Don't cook her a meal, take her to a restaurant instead. Don't make music, buy music. Riding the bullet train from Tokyo to Osaka through hundreds of miles of desolate industrial landscape, I have to remind myself that the people who live there are richer than most Europeans. They don't always think so. In one survey the Japanese who were questioned reckoned that they had, in reality, a lower quality of life than every European country except for Portugal.

Adding up all the financial transactions by all the companies and institutions in a country, converting it to dollars and dividing it by the number of people in a society does not tell you how comfortable they are. Cold climates have to spend more money than warm ones. In Britain, if you want hot sun or cold snow it gets expensive. In Italy they have these things for nothing. Poor Italians! Income is not equally distributed between people or between organisations and people; Japan keeps more of its money inside its organisations than Britain does; ageing societies spend more than young societies and have fewer people. They therefore look richer, but can feel poorer.

When the IMF used Purchasing Power Parities (PPPs), instead of the normal market exchange rates, to compare the output of countries, they found that China, for instance, went from $370 per head to $2,460, and India from $275 to $1,255. A little money goes a long way in China and India. PPPs reflect that fact. The $370 was clearly unreal when 70 per cent of Chinese urban households have colour television and 80 per cent have washing-machines. When the IMF added up the output of all the countries of the developing world on this new basis, they discovered that the developing world's share of the world's output had jumped from 18 per cent to 34 per cent, and that of the industrial world had fallen from 73 per cent to 54 per cent. China, in fact, because of all the hundreds of millions of its people, is, on a PPP basis, the second-richest country in the world, after the USA and above Japan. The way you see things depends on the way you count them.

There have also been attempts to add some of the invisibles to the visible. Most countries make an estimate of their informal, 'black', economies; one or two add this estimate into their national income. In one year, 1987, Italy jumped over Britain in the international league table when she did this, adding 18 per cent to her GNP in one keystroke of a statistician's computer. No one has yet added home work or gift work to the money numbers but the time might not be far off. It would be a painless way to get rich quicker and would benefit Britain with her long tradition of voluntary work. We should, therefore, remember the principle that no one set of numbers can ever serve all purposes. What we need are two sets of national accounts, one which records the money transactions and one which lists all the other indicators of life.

Counting Invisibles

This second list would include health and death statistics, infant mortality, age of death, cause of death. It would include education numbers, employment numbers and statistics on other forms of work; there would be details of housing, of environmental indicators such as carbon-dioxide emissions, deforestation and energy use and of more subjective indicators such as people's feelings about their quality of life. All these numbers currently exist in most countries. In Britain, many of them are published annually in documents such as *Social Trends* which has its equivalent elsewhere. What we need is not a new system of national accounts so much as another companion set of national statistics which can be compared year by year and country by country. The UN's International Comparison Program, which attempts to do some of this, may yet turn out to be one of the more important of its initiatives.

The statistics collected, we need to give these numbers the same sort of public prominence as we give to the money figures. They should, for instance, be presented as an annual review to every parliament or congress, be debated and discussed in the media and contrasted, for celebration or lament, with the statistics of other countries. Over time they would provide a set of benchmarks for a civilised society, to run alongside the national-income figures. Both sets of figures are necessary, if what we count affects the way we behave, and if we want a more balanced and just society.

Practical Action

We could start the reforms by making the national-income accounts a little more honest. Governments run the country on a cash-flow basis, 'money in' versus 'money out' each year. The difference is the deficit, or what the British quaintly call the Public Sector Borrowing Requirement. This allows them to take no account of the difference between an investment and an expenditure – they are both outgoings, even though the investment may save money in the future whereas the expenditure is gone for good. Education, therefore, is always a cost and never an investment. The cash-flow convention allows them to sell assets and call it revenue, even though it will never be repeated. The same convention allows them to treat bonanzas, such as Britain's North Sea oil production in the 1980s, which was never going to last very long, as an addition to revenue rather than the equivalent of Aunt Agatha's legacy, a one-off bounty, something to be invested in one's future.

The result is to distort priorities. There is no incentive to think long-term. There is no way to trade an expenditure today against savings or benefits in the future. There is no need to take account of future liabilities piling up; the costs of not maintaining roads and railways, or of the pension liabilities that accrue for each worker in the unfunded state-pension scheme. If we behaved that way in our own lives we should never buy a house, we would run our cars until they fell apart, and we would spend the minimum on our children's education, because the long-term future would always come second to paying the bills. In our private lives we get round that problem by turning large lumps of investment, such as a house, into smaller streams of expenditure, by means of a loan or a mortgage. If we are wise, we borrow only to finance investment not to cover the monthly bills. Government muddles them up. No business would want to behave like that, nor would it be allowed to. Politicians have always and consistently resisted the pressure to do their accounts in a proper 'businesslike' way, arguing that it would tie their hands unnecessarily and that, one way or another, they have to finance both the running deficit and the capital expenditure by borrowing, so why separate them out artificially?

One country, however, has made an attempt to be businesslike and to present a proper balance. New Zealand produced a national balance sheet for the first time in 1991. It revealed that its assets, state companies, roads, lands and buildings, financial reserves and investments totalled NZ$14.4 billion less than its liabilities, by which it meant its borrowing at home and abroad and its pension liabilities. Technically, the country was bankrupt. Realistically, it means that future taxpayers will have to pay for the relative profligacy of their predecessors. New Zealand will continue to publish its old cash-flow accounts but the new 'business' accounts will help to show how well the present and the future are balanced. The *Economist* calculated that, using these sort of figures, the net worth of New Zealand Inc. had deteriorated by $12 billion dollars over the past 20 years. By not counting, you could say, the New Zealanders had mortgaged their future. The present system of government accounting in other countries allows expediency to flourish because no one knows the true costs. Better counting would allow a more informed debate about the longer-term balance of priorities and would bring the issue of national purpose and direction to the surface.

A Scorecard for Business

Companies may have better balance sheets than governments but they, too, have a long way to go. These things do not normally count or get communicated:

The intellectual assets of the company (including their brands, their patents, their skill base)
- Their expenditure on the enhancement of these assets, including R. & D., training and development
- The introduction of new products or services
- Employee morale and productivity

The customer
- Quality of goods and service
- Customer satisfaction

The environment

Investment and expenditure on environmental control and improvement
Expenditure on community work
Investment in the community

These things are difficult to count and in themselves they mean nothing. It is only when you start to compare last year with this year, or your company with your competitors that the numbers get interesting. Comparison provides the benchmarks. Without any sort of numbers, however, the cash-flow numbers are the ones that count. It is hard, then, to know whether the business is in proper balance, whether the future is receiving the right resources and whether the stakeholders' requirements are in balance. If, as an investor or a customer, you are betting on the intellectual property of the concern you will need more than historical money numbers.

William Reilly, when administrator of America's Environmental Protection Agency, was asked what the Eastern Europeans should do first in the long haul to clean up the massive pollution in their countries. He replied:

> My answer is to begin with the disclosure of emissions. Require that the data be published in the local newspapers. Then support a healthy non-governmental, environmental movement. At that point a fascinating dynamic will begin to occur: the community will interact with plant managers, workers and government to reduce pollution levels. Such is the power of information.

Counting it makes it visible, and counting makes it count.

IBM now measures each of its 'Baby Blues' on seven parameters: four financial numbers (revenue growth, profit, return on assets and cash flow) and three new measures (customer satisfaction, quality and employee morale). In Britain, Dr David Budworth is exploring the concept of an 'innovation ratio', relating the amount spent on innovation (research and development, training, and the development of brands) to the value added by the company. Others are looking for ways to measure a company's 'knowledge bank'. Some already do it anyway, in their published accounts under the heading of 'intangible assets'. The trouble is that you have to read the very small print in the Notes to the Accounts to find out what this means, and it will mean different things to different firms, measured in different ways. To a publisher it can mean the publishing rights which they hold. To WPP, the communications firm, it meant the brands of their two big advertising-agencies, J. Walter Thompson and Ogilvie Benson Mather. They did not say how they had valued them, but it was, at least, a recognition that intellectual capital had a value.

Putting estimates of intellectual assets on the balance sheet may, however, only confuse matters. If we give them money numbers we shall be trying, once again, to use one set of numbers to count different sorts of things. Just as we use different measures for liquids and solids, so we should happily use different measures for each stakeholder. For the environment, for instance, the United Nations initiative suggests that each organisation should include in its annual report:

- the organisation's environmental policy
- the capitalisation of environmental expenditures
- any environmental liabilities, such as bringing the organisation into line with new regulations
- disclosure of other anticipated environmental expenditure

The Pearce Report for the British Government would, if it were ever acted upon, require organisations to disclose their man-made, natural and critical capital assets and the costs of maintaining them. Other possibilities are a mandatory environmental audit to monitor observance with environmental standards or a full energy-accounting system.

These are just a taste of the numbers being looked at. As a result of these new numbers, the environment is in danger of consuming more forests through measuring the forests it is saving. The environmental-cleaning business is now thought to be worth over $60 billion in the US alone, and growing. The bigger the problem, the bigger the business. Germany is estimated to have nearly 40 per cent of the Eastern European environmental market. If you turn the problem into a business, the numbers will appear.

Consumer needs are another growing business. Sensible businesses recognise that contented customers are faithful customers and will run surveys, collect data and analyse repeat buys. As more organisations realise that they are businesses even if they don't have shareholders, the practice is spreading. Britain's hospitals have started a customer-response record, to keep track of how satisfied their patients feel with their treatment. So have prisons. Neither hospitals nor prisons want their customers to come back, so they have to turn the repeat-business statistics upside-down. Big numbers are bad. Comparing their non-repeat business with their peer institutions across the country is a valuable benchmark. They have, at the very best, to explain why they are different. The different numbers set different agendas.

For a proper balance, organisations need an audit of their relationships with all their stakeholders, even if some of the details should remain confidential. There would be no damage in publishing details of a firm's involvement with the community, or its investment in its people. One sign of the changing priorities of the times is the number of firms who see some competitive advantage in advertising their activities in this field instead of merely reporting them.

'Join us,' advertised one accounting firm, ' and we will invest at least 10 per cent of your salary in your development each year.' 'We promise every employee the chance to work one day a month in the community at our cost,' declared another. If it is seen as a business opportunity, the numbers emerge. Details of intellectual property or supplier relationships may be more private, although audited general ratios such as an innovation ratio would give nothing away to competitors but would be some indication of the level of investment in the long-term capacity of the organisation. These could and should be public. It should be good business to publish good ratios because shareholders might be impressed.

Many boards of large companies now have separate audit committees for social policy, for ethics, for remuneration and for the environment. Some, such as ICI in Britain, publish a separate environmental report alongside their annual report. This must be a good start. It is another recognition of the hexagon, of the variety of interest groups and of the multiple contradictions involved in charting the path of a business. It will be better still when we can make the data public, on a standard basis. A social audit is required in France as part of the annual report. My guess is that something similar will soon be required of all European companies. They will resent the bureaucracy and the costs involved, they will chafe at the restrictions on their freedom of action, but by counting the invisibles they will better balance the present against the future and the interest groups against each other. Sometimes, it seems, we have to be forced to be sensible. Once the playing-field is level and the rules are known, the game can start.

A Personal Scorecard

'How much money do you earn?' I used to ask my friends in my competitive days. It seemed the best way, then, of comparing progress in life, after aiming off for the fact that I was then an oil executive while some of them were bustling bankers and others exhausted young doctors. I was brought up short by one who replied, 'Enough.' 'What do you mean – enough?' I asked. 'What I say – enough. I work out what I need and that's what I make sure I earn. Why bother to make more? How much sugar do you buy in a year?' he turned and asked me. 'I have no idea,' I said. 'But I bet that there's always sugar in your house when you need it. Money is like sugar, no point in hoarding it, it usually goes bad, or you have to make quite unnecessary cakes to use it up.'

Crazy man, I thought; but as I grew older I realised the sense in what my friend had said. He was never rich but, as he said, 'there was always sugar in the house' and he seemed much less harassed than the rest of us. But then he knew what he wanted out of life. He wasn't using money as a substitute for uncertainty. In a time of materialism most of the numbers are financial ones. The higher we score, the better we do, apparently, but like the sugar we then have to go out and spend it, often on the equivalent of cakes which we don't really need. In the recession, many couples found that they were unable to trade up in the housing market as their families grew along with their income, because they could find no buyer for their current home. 'I was frustrated at first,' one of them told me, 'but then we said – we've been very happy in this place; maybe it's a bit small and a bit un-smart, but that doesn't really matter. Let's go on enjoying what we've got and take the hassle out of life. We've got enough.'

Money is seldom the measure of much, once you have enough. It is only the core of our personal doughnuts. I cannot be the only person to wonder what those people who are paid one million pounds or dollars in a year do with it all. Sugar goes bad if you don't use it. Money isn't even necessarily the sign of success. In Britain the quickest way to get rich is to fail at the top. Sign a three-year contract and then fail in the first six months, walk away with a million pounds for six months' work. As I grew older I realised that my friends were not that impressed by those who had riches, as long as 'there was sugar in the house'.

Some Hints of Change

If money is not the measure of all things, how do we put numbers on the other things – a walk in beautiful countryside, artistic expression, the love of a family, the joy of teaching, watching someone get well, the thrill of discovery, the satisfaction of a job well done, the delight of friends? We all know the taste of such things but find it hard to call them success. We need to find a way to list them even if we can't count them. We can learn a lot from our children, particularly when they grow into adults. I once asked my 24-year-old daughter what she thought she was doing with her life, dabbling in this and that, travelling, adventuring, socialising. 'When', I said crossly, 'are you going to find a proper career, make a serious contribution to this world?' She looked at me, a little pityingly, I thought. 'There are many people,' she said, 'who rely on me for comfort and for help, who use me as their home. I learn something new each day, I laugh with someone every day and I cook for someone almost every night. Oh, and I do no one any harm. I don't think that's bad for 24.' I went away, wondering how long it would be before I could say the same.

Laurence Shames, in his book *The Hunger for More: Searching for Values in an Age of Greed*, puts the dilemma this way:

> The frontier . . . is what has shaped the American way of doing things and the American sense of what's worth doing . . .

> More money, more tokens of success – there will always be people for whom these are adequate goals, but those people are no longer setting the tone for all of us. There is a new sort of *more* at hand: more appreciation of good things beyond the marketplace, more insistence on fairness, more attention to purpose, more determination truly to choose a life, and not a lifestyle, for oneself. Dare we suggest that these new forms of *more* comprise a species of frontier?

Measuring more, he goes on to say, is easy, measuring better is hard.

Mickey Kaus, in America, shares my worries of a society increasingly divided because money is the measure of so much. He would like to take more things out of the market so that it did not matter whether you were rich or poor. The National Health Service in Britain does not differentiate between rich and poor. I am always relieved to land at Heathrow Airport after a trip abroad because I know that I can now afford to be seriously ill. Kaus would add the draft, or National Service, all schools and colleges, parks and class-integrated housing. I would add all public transport. His dream that there would ultimately be a society which regarded a janitor as being 'just as good as a banker because he works as hard' must be unrealistic but the notion of taking as many things out of the money economy has much to commend it, expensive though it would be in taxes.

My own hope lies more in the denizens of the Third Age of life. They will not, most of them, have much chance to add more sugar to their store. Enough of that will have to be enough for them. They will then begin to find that there are satisfactions and achievements which cannot be measured by money, that gift work and study work and home work of all types can be richly rewarding. Because there are going to be a lot of such people they will be noticed. They will not be old, as we used to think of old, they will not be retired in the way their parents were retired, and most of them will not be poor. They will establish new sets of case law, some new models for success, some new numbers. How many young people have you coached this year, they may enquire, how many paintings finished, gardens planted, books read or even written? How many school trips did you organise, how many patients driven to the hospital, what moments of quiet beauty did you catch, which fireside chats have you treasured, what special meals, what letters written or photographs framed, friends counselled, feuds settled, loves kindled?

Imagine one of those lambent June evenings in England, when the sun doesn't set until after nine, and the air is still and scented – I was walking along the river-bank in Cambridge, with the immaculate lawns and the hauntingly beautiful chapel of King's College in front of me when a snatch of treble voices in a choir floated over the trees and a young American couple stopped, entranced. 'Remember this, honey,' she said, 'remember it always. This is quality time.' If we are going to find a better balance in our lives and a better justice in our societies, we need to find more examples of quality time, make them more accessible to more people, and make them count. We can do that by celebrating them more, for fashion is a powerful agent of change.

Part Four:
The Search for Meaning
Making Sense of Paradox

The Three Senses

The White Stone

'What is the point of it all?' my friend asked. 'Why should we struggle with these second curves, doughnuts and compromises? In the end, isn't life just a sick joke?' I knew how she felt. I had recently watched my wife's mother dying. One month she was the twinkly, irascible old lady at the heart of the family; the next she was a grey, emaciated shape in a hospital bed, hardly able to smile, let alone talk. After that, nothing – a small pile of ashes in an urn. Could this be all there was to it?

For some there is no point. Chekhov said: 'You ask me what life is? It is like asking what a carrot is. A carrot is a carrot.' Maybe, as Gertrude Stein once said of Oakland, California, 'There is no there there.' We are all accidents in the evolutionary chain. We can lie back and enjoy it, or we can occupy ourselves, as scientists do, in trying to understand more about what is going on. There is, however, nothing which we can do to alter it, even when we understand it. We can only play with it. Man is as the smallest piece of dust in the universe. Descartes thought that animals were machines. Some biologists see no reason to think that humans are any different from animals.

This 'myth of science' so frightened Allan Bloom, when he saw the effect it was having on modern American youth, that he wrote his best-selling book *The Closing of the American Mind*. American college students, he observed, were not only lifeless and ignorant, they were reluctant to offer or to hold any opinions at all. People who thought that they were right in the past did terrible things as a result, therefore it is best to have no opinions at all. The only true knowledge is science. Everything else is wishful thinking. From that it follows that it is wrong to take a position on anything, worse still to try to impose your wishes on your bit of the world. A passive voyeurism will have to suffice, preferably uncritical and politically correct because it would be wrong to suggest that any one way of life was superior to another.

That way lies a moral vacuum, where nothing is right and nothing really wrong. That way also lies inertia, no second curve, with compromises made for the wrong reasons – for a quiet life, not for justice or for progress. Immanuel Kant, then an obscure lecturer in philosophy in an obscure Prussian town in the eighteenth century, disagreed with Descartes. He sat down, wrote his *Critique of Pure Reason* and made the world stop and think, because in it he offered an alternative to the scientific arrogance which then, as maybe now, held sway. Man, he maintained, was not a means to an end, he was himself the end. Man's life was driven and shaped by a moral pressure, which came from within. There is something about the human condition which implies something religious, however we want to describe it. It is God in the human soul, Kant said, not God the architect of the scientific universe, who makes sense of who we are. 'How do you know this?' he was asked. 'Because of the moral force within me,' he replied.

It is as good an answer as I know. Faith has no reasons. If there were reasons, or logic, there would be no need of faith. I cannot prove that there is a point to our existence. I agree with the philosopher Ludwig Wittgenstein, who said that, 'Even when all the possible scientific questions have been answered, the problems of life remain completely untouched.' I also agree with John Updike, who said that existence felt like ecstasy, even if we were not able to describe it or define it. Even if it is a conceit, we feel that we know that we have something like a soul, that we matter, and that we are in some small way unique. There is a haunting passage in the Book of Revelation in the Bible: 'to anyone who prevails, the Spirit says, I will give a white stone, on which is written a new name which no one knows except he who receives it.' I keep a white stone on my desk as a reminder of my uniqueness. Even if there is no point, even if it is all a game of science, we must still believe that there is a point. If we don't believe that, there will be no reason to do anything, believe anything, change anything. The world would then be at the mercy of all those who did believe that they could change things. It is a risk we cannot run.

To find that point, that reason for our doing and our being, it helps to build on three senses – **a sense of continuity**, a **sense of connection** and a **sense of direction**. Without these senses we can feel disoriented, adrift and rudderless. The world is going to be a confusing place for the next few decades. We shall need all the help that we can find to recognise our place and role in it. These senses are the best antidote I know to the feelings of impotence which rapid change induces in us all.

14 A Sense of Continuity

Cathedral Philosophy

A few years before he died my father gave me a dirty brown envelope. 'I'm never going to get round to this now,' he said, 'so you had better have it.' It was a collection of old family papers including one of those family trees going back two or three hundred years. I looked at it and noticed that there was an actual namesake of mine a few generations back, one Charles Handy, who was born in 1765 and died in 1836. He married and had four children of whom two died early. One of the other two was my great-great-grandfather. That was all I knew. The paper told me nothing about where that first Charles Handy lived, what he did, how he looked, how rich he was, or whether he was a nice sort of chap or not – nothing. 'Is that what will become of me?' I wondered, a name on a family tree to be opened 100 years hence by someone I know not of. Is that what it's all about? Why, then, I might as well make merry and die.

It was then that I, by chance, came across the last few verses of the Book of Ruth in the Bible. These are hardly famous verses: they consist entirely of a list of names – 'Pharoz begat Hezron,' it goes, 'and Hezron begat Ram, and Ram begat Amminadab . . .' and so on for another six names, and then, conclusively, 'and Jesse begat David'. But that was the point; David was the point. He was the great king of the Jews, and the link, ultimately, with Jesus. Without all those other people there would have been no David, they were the essential links in the chain. Without that eighteenth-century Charles Handy, I would not be here today.

I realised, then, that I should not have been so arrogant as to think that it was up to me to make a great contribution to the future. Great, if that was how it worked out; but my main task was to ensure the continuity, not just of my family but of the things that I believed in. Forget the literal meaning of 'begetting', treat it metaphorically. It can apply to institutions and ideas as much as to kith and kin. We are links in a chain; it is up to us to keep things going because who knows which generation will be the one to make the big difference. David came nine generations after Ram.

The definition of a wise person in the Book of Proverbs reinforces the importance of continuity – he should ensure 'that there should be an inheritance for his children's children'. Jonathan Rauch, in his insightful book about Japan, *The Outnation*, describes his meeting with Yasunari Hirata, who started a business in 1946 making pushcarts and baby carriages, but is now making industrial robots. Talking of his job, Hirata says, 'I see the company as an infinitely growing child. I will die, but it continues to live, and my responsibility is to see to that. And I want to continue to build better and better robots.' Jonathan Rauch goes on to say that the word 'profit' would not have passed Hirata's lips if he, Rauch, had not brought it up. 'Yasunari Hirata was plainly not particularly interested in profits – not, I mean, in the sense of *taking profits*. He did not live lavishly and he seemed more concerned with immortality than with money.' He was expressing something which Richard Hooker was saying in England at the end of the sixteenth century: 'The Act of a Publick Society, of men done five hundred years sithence, standeth as theirs, who are presently of the same Society, because Corporations are immortal.' These ideas are of long standing.

John Rawls, the philosopher of justice, says, 'Each generation must not only preserve the gains of civilization and culture, and maintain intact those just institutions that may have been established, but it must also put in place a suitable sum of real capital appreciation.' Long before him, Edmund Burke, writing about the French revolution, said: 'Society is indeed a contract . . . not only between those who are living but between those who are living, those who are dead and those who are to be born.'

It is cathedral philosophy, the thinking behind the people who designed and built the great cathedrals, knowing that they would never live long enough to see them finished. The new cathedrals will not be of stone and glass, but of brains and wits. They will take equally long to build and we who must start the building may not live to see the conclusion. They is why we need to look beyond the grave and beyond our generation. It is hard to believe that we will make the sacrifices involved unless we can believe in the long-term existence of our little local world and of the bigger global one. We should, however, remember that there is no need for that continued existence to be in the same form as it is at present. The second curve is different from the first; there has to be change to be continuity. Yasunari Hirata started with push-chairs and moved on to industrial robots. We need to have faith in the future to make sense of the present.

How Big and Far Should We Think?

Some would say that even life itself is now under threat, that Malthus' fears, two centuries ago, that the world would not have the resources to feed its peoples, are now coming true. The numbers do indeed look frightening, and the arguments of people like Paul Kennedy, in his recent book *Preparing for the Twenty-First Century* or Edward Wilson with *The Diversity of Life* are horribly convincing, but the end is not the end if we don't wish it to be. We may need to adopt the sort of measure that Lester Thurow suggests, paying rents to the Third World for their forests. This payment might encourage us to develop the millions of different life forms which Wilson identifies in his book. One way or another, we shall need to have enough faith and interest in the continuity of the world and its peoples to give up some of our present wealth for the unseen benefits of people whom we will never know. It will need a bigger sense of continuity and of cathedral thinking than now exists.

Some people have that sense in their culture. Charles Hampden-Turner argues that Americans see time sequentially, as a straight line, whereas the Oriental races see it as a loop. To many Westerners, time, he says, is a running reaper, waving his sickle. But in the East, time comes round again and again to seek people's engagement in new opportunities. If you take the running-reaper view, then there is no time to lose. Things must be completed before time runs out. In the loop view of things, time never runs out. Therefore what you want to create are self-renewing systems, systems which will still be in place when time, your friend, comes round again, even if you are not there.

We could look at Europe in this light. In one way or another, Europe has to become more integrated. Its separate countries are too small to go it alone in a world of eight billion people. But for Europe to succeed, it must be more than just a convenient arrangement for the exchange of goods. It must, in the end, be a federal unity with all that federalism implies, twin citizenship, separation of powers and proper subsidiarity. Federalism also requires a common law and, ultimately, a common currency. At the moment, the unequal economies of Europe adjust their currencies when they have to, in order to restore their competitiveness and to allow them to sell their goods. Once there is a common currency, devaluation is no longer possible. Europe will then have to do what individual countries do to equalise their regions – make grants, loans and tax concessions to help the weaker catch up with the stronger. For Europe to have the cash to do that adequately, we shall probably have to increase the tax which we all pay to the centre, possibly by a factor of seven or eight. No voters will be prepared to do that unless their sense of history, of continuity backwards and forwards, allows them to see themselves as an integral and continuing part of this place called Europe, a place which has been a part of their heritage and a place which will contain the future of their great-great-grandchildren. There will be no short-term rewards for this sort of sacrifice, no way of justifying it within the lifetime of a parliament.

In a business, quarterly reports and an average lifespan of 40 years for big companies tend to put immortality on the back burner in most boardrooms. Boardrooms always want numbers, but trust no numbers beyond four or five years. It is always safer to put one's money on deposit than to risk building some cathedral of new enterprise. Only family businesses have the urge to think beyond the grave, and even then probably not beyond three generations of graves. The pressures to do it in your own lifetime seem to be mounting, even though the institutions who currently own our public companies are themselves supposed to have a continuing existence, independent of the people who run them. We need to re-emphasise that institutions can be immortal even if we are not. The Mitsui Corporation and my old Oxford college are both over 600 years old, both still going strong and thinking far. You only look ahead as far as you can look back.

Invited to help one large bank think out a statement of its visions and values, I looked, to start with, for their understanding of their purpose. 'Why do you exist?' I asked. 'To make our shareholders seriously rich,' they replied. Their shareholders were mainly other banks, insurance companies and pension funds. Had any of these, I asked, ever made that request of them, or defined for them what their expectations might be? It appeared not. The chairman added that, after announcing their first-ever annual loss, some years back, he had thought it right to call on their principal shareholder, an insurance company, to explain the situation. 'They were noticeably uninterested,' he said. 'They seemed to assume that we would be around for ever and that this was only a temporary hitch which we would put right.' Maybe they were right, I suggested; maybe they wanted to keep their money in a place where it could stay for ever. 'I think that they are very unusual,' he replied, disposing of that suggestion. Immortality as a concept can be frightening. Continuity can, however, be a useful and less scary compromise.

On a more domestic scale, we have to worry about the grandchildren. Now that one in every 2.3 marriages in Britain is ending in divorce, it is not clear who will be thinking about the grandchildren when there will be an abundance of unaccepted step-grandchildren or, looking at it the other way up, many children with grandparents whom they have never known. It is hard to plant trees in your garden if you do not know who will be around to look upon them when they are fully grown. More formally, the concept of justice between the generations will be harder to maintain when the generation after next is semi-detached. Families used to last for ever. It made sense to say 'when I am gone', knowing that much would continue even if subject to change. If families become things of temporary convenience, or inconvenience, time will indeed run out.

My hopes are fragile ones. We should not, for a start, underrate the power of the millennium idea. It will encourage people to look backwards and forwards farther than they have ever done before. The heritage movement is also gathering pace. We don't pull down things so often but refurbish them instead, turning our docks, warehouses and factories into new uses. That is usefully symbolic.

RISC, the International Research Institute for Social Change, reports, from its recent surveys, that 'we can witness an increasing sense of responsibility towards the flux of history, including a greater recognition of the importance of both past and future generations'. That is encouraging. The environmental campaigns, and particularly the idea of Gaia, of earth as a self-renewing system, take us on huge leaps back into history and forward into the future.

If companies, as I hope, rediscover the virtues of membership for their chosen ones, if job-hopping becomes more perilous and if shareholders wield less power, then the corporate world may see a desire for permanence creep in again. Corporate leaders do wield influence, so when 48 executives from the largest corporations in the world come together to form the Business Council for Sustainable Development, governments and others will listen. Carl Hahn, as chairman of Volkswagen a member of that group, wrote in its report: 'If we think of the future – a central part of the obligation to rising generations – we must adopt the cyclical approach on which the whole of nature is based.' In the market-place, fashion, that god of the merchandisers, may be losing some adherents, or rather, the new fashion may be to make do with what you have, or to choose what suits you and not your neighbours' fancy. Home-made, second-hand, good not flashy quality, the stuff that lasts, may become the style.

Now that people live longer and four-generation families, with great-grandchildren, become more common, the idea of continuity in a family may yet re-emerge, when today's footloose parents become tomorrow's grandparents and realise what they are missing – a stake in the future. Because we look forward only as far as we can look back in life, this realisation can only come late in life. The rights of grandparents would then become an issue, with children effectively adopted by their grandparents irrespective of their parents' new arrangements.

Without a sense of continuity there is no point in sacrificing any of the present for the future.

15 A Sense of Connection

We were not meant to stand alone. We need to belong – to something or someone. Only where there is a mutual commitment will you find people prepared to deny themselves for the good of others. We, however, in our belief in liberalism and individualism, are wary of commitments. We look suspiciously at words like 'loyalty' and 'duty' and 'obligation'. Independence, whether we seek it or not, is being thrust upon us. 'Modern society knows no neighbours,' said Disraeli more than a century ago, and it has been no different since. Loneliness may be the real disease of the next century, as we live alone, work alone and play alone, insulated by our modem, our Walkman or our television. The Italians may be wise to use the same word for both alone and lonely, for the first ultimately implies the second. It is no longer clear where we connect or to what we belong. If, however, we belong to nothing, the point of any striving is hard to see.

More crucially, perhaps, if we belong to nothing, there is no reason to make sacrifices for other people. Duty and conscience have no meaning if there is no sense of commitment to others, and of others to us. 'Think of a person,' said Rawls, 'without any sense of justice. He would be without any ties of affection, friendship or mutual trust, incapable of resentment or indignation. He would queue-barge if he could get away with it and expect everyone else to do the same. He would be less than human.' The interesting question is, then, not why some of us are criminals, but why more of us are not, in a world where so many of the connections which underlie that sense of justice are breaking down.

The workplace has been the central community in the lives of many in this century. Lewis Mumford, extolling the virtues of the monastic community, said, 'True leisure is not freedom *from* work but freedom *in* work, and, along with that, the time to converse, to ruminate, to contemplate the meaning of life.' Modern work does not provide too many of those opportunities, even for those in the core. Even so, the lament of the prematurely retired is usually the loss of community, while the loneliness of the long-distance teleworker is well documented.

If the idea of membership becomes more prevalent for those in the core of institutions, then the workplace will remain a central point of connection for many. That 'many', however, is likely to be less than half of the workforce and less than a third of all adults. It is a connection, however, which could be a very isolating one, consuming all its members' time and energy, and insulating them from the surrounding society, unless more attention is paid to the corporate contract with the other stakeholders and more trouble taken to chunk time sensibly.

The New Ghettoes

This has been one unintended consequence of the organisation society; the key place of the organisation in our lives removed from many of us the need to belong to anywhere other than our workplace. As a result, when we leave it we have nowhere. We also substituted the homogeneous communities, which our work provided, for the mixed communities of the old neighbourhoods. We replaced the community of place with the community of common interest. When you do that, there is no longer any need to think of sacrificing anything for your new neighbour because your neighbour is in the same position as you. If we then compound that by turning our communities of place, where we choose to live, into equally homogeneous zones, we shall never need to see, meet or pay heed to anyone different from ourselves.

There were, in 1989, 130,000 community associations in the United States, according to the Community Associations Institute, helping to administer the lives of 30 million Americans, one out of every eight. Some of these are just small condominiums, but 80 per cent of them own land as well and have an average of 543 dwelling-units. More are on their way and are getting more homogeneous; one new development in Newport Beach has even put a limit on the size of residents' dogs.

There is also the Leisure Hills development in Laguna Beach in California – 21,000 people with their own taxes, security force, television station and 12 bus routes. Guards at its gates check the identity of all visitors. It, and the other ghettoes like it, are the equivalent of the walled cities of medieval Italy. They provide their inmates with the security and peace of mind which they cannot find in the mixed community. They should remember, however, said the *Economist* when reporting this, that those walled cities of Italy were the source and cause of endless wars.

The new ghetto communities are too small and too like-minded to be the basis of any new balance of society. They are only connected to themselves. On the other hand the nation is too big and amorphous a concept to count as a connection. We are not easily going to be persuaded to make sacrifices for people whom we never see, to pay to clean streets we never walk, or to mend sewers we shall never use. The rich of Surrey may feel sympathy for the poor of Tyneside in the North but they will not put too much of their money their way, because they will never see the results. The organisation society has gradually become the ghetto society, ghettoes of the rich and ghettoes of the poor. We should remember that it was the enclosure acts of Henry VIII in England, done in order to get better productivity from the better farmers, which forced the poor into poverty and into a new underclass. We need a community which is large enough to be a mixture and small enough to be visible to all its inhabitants. We need to return to the city state or at least to the township.

Civic Pride

There used, in Britain, to be a thing called civic pride. Town hall would compete with town hall in magnificence and in achievement. Central government down the decades has progressively stripped the cities and the towns of their powers, distrustful of how they used those powers and, in some cases, undoubtedly abused them. All that is now left in Britain of this tradition is the city football club.

There was a time when the municipal universities were the pride of the city fathers; businesses competed to see their names inscribed above some new hall or lecture theatre, their sons and daughters studied there, married there, lived and worked there. In the 1950s the British government decided, in a spirit of liberalism, that they would give fees and maintenance grants to allow students to study at any university in the land. The city universities instantly became national universities. They lost their local identity and their local patronage; students roamed freely, made their contacts and their roots far from the city of their birth. It was done in the name of freedom or, say some, because the universities of Oxford and Cambridge wanted to have the pick of all the land. Whichever, it was one more blow against the city state.

The city may, however, be on the rise again. Europe is fast becoming a Europe of the cities. Manchester competes with Barcelona, which competes with Munich, in business and in sport. Airlines fly from city to city, not only from capital to capital. Cities twin with cities. It makes good sense. We can identify and connect with a city, even if we only live in its hinterland. The castle, the cathedral spire and even its tower blocks are a visible reminder of its presence. The city is a community on a human scale, the nation state is not. The only people who wave the Union Jack these days, or the French Tricolor for that matter, are drunken sports fans. As the middle orders disappear in a more integrated Europe, the city is replacing the nation state as the focus of our identity and our way of connecting with society.

But our cities are a mess. They represent the extremes of riches and poverty, of affluence hobnobbing with squalor. They look to be an unlikely basis for community. It is for that reason that they should be given back the responsibility for their futures. In Britain the bulk of the income of the cities comes from central government. The cities are only the delivery agents. Proper subsidiarity requires that they have the right to decide on their own priorities and are given both the authority and the means to deliver these.

Cities, together with their hinterland, are the best basis for the Chinese contracts which are required of a fair society. Only in that size of community will it be possible to harness the talents and the money of the more successful in order to provide the investment in the infrastructure and the help for the less fortunate. Those who give will be able to see the results of their giving or their taxes. In a city you can make some difference in your spare time. At the national level making a difference is a whole career. Giving the cities more responsibility, however, also means giving them the right to raise the money to deliver that responsibility. There lies the rub, because central government in all countries, save perhaps Canada, likes to have the first and largest bite of the tax take. It will require a radical central government to give up its fiscal control to this extent.

America's cities are farther down the road than most European ones, but everywhere there are some encouraging signs. The idea of an annual European City of Culture has begun to catch the imagination. Cities, rather than countries, are promoting themselves as centres for tourism and development. They compete for heritage and environmental awards. Cairo, with its 15 million population, recently received a United Nations award for recycling its rubbish, proving that size need be no drawback to civic pride. Only London, amongst the world's large cities, has no government of its own, no centre for civic pride.

The hard truth is that federations should be both small and big, with the inevitable result that the middle levels fade. Europe will one day be a federation of cities in all but name. When it is, then it will mean more to each of us to be a European because every city will need to rely on Europe and its connections far more than they do today. Reciprocity, too, is easier when there are many different players. It is easier for Glasgow to exchange people or projects with Oporto than with Birmingham, because it is less directly competitive. It is also more fun for both. The idea of twin citizenship is, it seems, easier to foster on a city basis than a national one.

I have long cherished the idea that at age 13 every child in Europe should spend a semester at a school and in a home in another European country. If done universally the only cost would be the travel. This is something which it would be hard to negotiate or to organise on a national basis. It could be much more fruitfully done on a city-to-city scheme. Nothing would do more to heighten a young person's sense of history and of a shared destiny than that connection.

For Europe, read North America, because surely, one day, geography and economic logic will combine to create a new and larger federation there, bringing in Mexico and Canada. That, too, will need to be broken down into cities and large towns rather than nations or even states, if the rich are ever going to be prepared to shell out for the poor. Twinning and swapping will then help to create a sense of shared history and shared destiny. Japan and South-East Asia may, one day, feel the same pulls to be both larger and, at the same time, smaller, if they are to compete with the growing force of China.

Most of the hope for cities, however, rests with the organisation, and particularly with the organisations of business. Businesses need the cities, they need the educational and cultural resources which the city offers in order to attract the quality of people whom they will need in their core. They will need the transport connections which cities alone can offer. They need the plethora of small service businesses and portfolio people who congregate on the edges of cities, in the new-style villages. They need the buzz of cities, the energy and excitement which you find there, the variety of life, the contacts and the political connections. Paradoxically, the trend at present is all the other way. Organisations are fleeing to the countryside or the suburbs, seduced by the dream of the office or factory in a garden, bringing the work to the workers rather than the other way round, relying on telecommunications to give them their link with the outside world. They are creating their own walled cities in the woods.

They may rethink. The new federal dispersed organisation does not have to have many people in any one place. Local work centres and tele-clubs can proliferate in the woods and the suburbs but the city needs part of their action and they need the city, for the stimulus of its connections. Only when the organisations return will they be prepared to invest in making these hubs of humanity civilised once again, because they have the clout, the spending-power and the leadership skills. Compromise may be the way forward. The symbolic analysts will increasingly work in more places than one, live in more places than one, not *rus in urbe*, the classicist's dream of the country in the town, but *rus* et *urbs*, for they will need them both. It is when the rich as well as the poor live in the cities and the towns again that there will be some chance that the rich will pay towards the education and the transport of the poor, because it will ultimately be in their interest to see their city better educated and therefore richer.

If organisations do not rethink and relocate we shall see the cities declining even faster, those of power and influence retreating still further into their walled villages and insulating themselves from anyone unlike themselves. That way there is no balance, less chance of sacrifice or compromise, less likelihood of turning paradox into progress. It is, therefore, heartening to know that the businesses of London are coming together to create a programme for London, that those of Birmingham are doing likewise, as are the leading citizens of Atlanta, Seattle, Barcelona, Seville, Glasgow and many more. 'Federations,' said Osborne, 'are the laboratories of democracy.' Some of our cities may yet, unexpectedly, find ways to reconnect us with our neighbours. Where they lead others may follow, as long as those who rule in the centre come to appreciate the benefits of federalism.

Virtual Cities

Great cities are made up of small villages. The truth is, as so often, that we need our village and our city. We need both the comfort of friends and the stimulus of strangers. We relax in the company of people like us, but we also need a connection with a bigger and wider society both to keep us from falling asleep, and to make us feel part of something bigger. Only then will conscience prevail over self-interest, and duty over comfort. Villages, even villages of like-minded and like-income people, in the midst of great cities would be a good basis for a fairer society. For most of us it will not happen like that. Our cities will not change fast enough. We must create our own virtual villages and cities, communities which one can describe and can visualise but which do not necessarily belong to any one place.

The family has always been one of our 'villages'. The traditional family is no longer so traditional, but there are still families. They may not be composed of conventional relationships, and there may be more step relations than blood relations, but there are still families. The extended family is now horizontal as well as vertical, covering a wider group of people of the same generation and offering a wider choice of soul mates than the narrower nuclear family of old. These new 'virtual families', which include close friends and partners as well as the blood relations, may well be more comfortable 'villages' than the older model. We should not despair of the family, but redefine it.

Work was another of those villages, often, as I have suggested, a ghetto unconnected with the world outside, but a comfortable way of connecting with like-minded colleagues. The spread of the minimalist organisation is making these connections more difficult, as more people move or are moved outside. Increasingly, the modern-day portfolio workers have to create their own 'virtual organisation' made up of clients and occasional partners or co-workers.

The virtual organisation can be glimpsed in the new 'clubs' for the independent portfolio workers. One of these clubs is provided by the intermediary employers, the employment agencies, who are the brokers for the portfolio workers, just as agents are for actors, writers and models. The intermediary employers, or agents, provide a reference point, a base and an ally, even if it is only at the end of a telephone line. There are also the networks of contacts which any independent soon builds up, the job clubs for the unemployed and the professional associations for those qualified to belong to them. My son, an independent actor, has both an agent and his 'filofax club' of contacts. These are his virtual organisation between engagements. One new development is the tele-club, a building designed to be used by the occasional tele-worker, offering cubicle space, receptionist services, food and drink, and all the necessary communications equipment.

Some of these facilities are hired by organisations for the use of their people, providing them with a local regional office. Others are for hire by individuals by the day, week or month. There are more up-market versions in the city, offering meeting and eating rooms for hire by strangers or more formal clubs which restrict their facilities to members. Hotels, airports and railway stations have seen the commercial potential in the new form of work, and have provided their own tele-clubs for travellers with a gap in their schedule. In time, these places may offer a form of temporary fellowship as well as their facilities, becoming a physical embodiment of the virtual work village.

There are other alternatives, other virtual villages. One study, delightfully called 'Organizing around Enthusiasms', discovered that there were 315 organisations in one Surrey suburb devoted to hobbies, interests, sports and other enthusiasms, all run by volunteers, all doughnuts with a core of organisers and a space full of subscribers and participants, all offering some sort of club to their members. They offer scope for activity, these places, not paid work, but they provide an opportunity to take comfort in the company of friends, a temporary village.

These virtual villages need then to be complemented by virtual cities, opportunities to meet and be challenged by strangers. Mickey Kaus advocates more use of what he calls 'Third Places', places like cinemas, churches, shopping-malls and other common meeting-arenas. More could be done to make these Third Places into opportunities for connecting with strangers. Too often, however, they are made up of lonely crowds. We cannot rely on these, but must do more to build our own connections with strangers. This is more difficult and more challenging but not impossible.

My brother-in-law left his full-time job in business at the start of his Third Age. He looked for part-time work for some four days a week. Four years later he thinks that he might manage to fit in an occasional day. He is far too busy to do more. He is a local magistrate, a governor of a local school, and a member of the parish council; he sits on various committees of the local judiciary and the police authority, runs the local gymkhana and part of the big agricultural shows in that part of the world. In place of one rather monotone business organisation he now belongs to a wide range of groupings. He sees sides of life which were probably undreamt of in his business office. Like him, we can use the new flexibility of work and life to make more connections than we would have in the days when the one organisation kept us fully occupied for most of life.

A portfolio of community activities, be they good works or good 'enthusiasms', is the antidote to ghetto life, be the ghetto a country village, a fenced city in California or a slum tenement in an inner city. You do not have to have my brother-in-law's background to serve the local community. Some of the best school governors, the best tribunal members and the best youth-leaders come from deprived areas. They have an understanding and a grip on reality which outsiders can only envy. Our community organisations promote the connections across the divides, both horizontal and vertical.

'There is no such thing as society,' said Margaret Thatcher, famously. She meant that individuals could not hide behind 'society', looking to it to provide for them or to protect them. 'Individualism,' said John Maynard Keynes, 'if it can be purged of its defects and abuses, is the best safeguard of individual liberty.' But the 'if' in Keynes's statement, the proviso, is important. The sense of connection which you can find in a mixed community is the best means of purging those defects. There can be a beneficial compromise between the individual and a community.

Society does exist and is necessary, but as a supplement to individualism not a substitute. Society is also an outlet for our contributions, a place to give to as well as get from.

We get more local as we get older and often feel the need to give something back to society, in time and expertise rather than money. As long as they don't hide away in their geriatric shrubberies and golf gardens with their size-restricted dogs, the Third Age people have a lot to give, but they will want to do it locally where they can see some of the results. We could see a new corps of para-professionals emerging in the community, individuals basically trained and competent enough to help with the Youth Service in the schools, in the hospitals and clinics as counsellors or drivers or attendants, in the dispensing of benefits and welfare, as researchers for projects or co-ordinators of volunteers. Some of this happens already. More could happen if civic pride were resurrected everywhere. We cannot wait for central government to give away its power, we have to do what we can without it. In the world ahead we shall increasingly have to make our own connections, our own virtual city and our own virtual village.

16 A Sense of Direction

The End of History

In the end, however, a sense of continuity and of mixed connections will not be enough to give point to any striving. Maybe nothing will. Francis Fukuyama, the author of *The End of History and the Last Man*, put it this way:

> The end of history will be a very sad time. The struggle for recognition, the willingness to risk one's life for a purely abstract goal, the worldwide ideological struggle that called forth daring, courage, imagination, and idealism, will be replaced by economic calculation, the endless solving of technological problems, environmental concerns, and the satisfaction of sophisticated consumer demands. In the post-historical period there will be neither art nor philosophy, just the perpetual caretaking of the museum of human history.

Fukuyama's argument is that liberal democracy, the tolerance which it brings with it and the affluence which made it possible, have removed the will to fight great causes. We are slumped in comfort. When we compete it is for the World Cup or gold medals. Such things do not bring forth great art or noble deeds, they don't stir the heart more than momentarily nor do they foster revolutions. Like dogs, if we are well fed we are content. When scientific and economic progress lead more and more societies into the contentment stage we shall see the end of history.

De Tocqueville saw it coming long ago, in America:

> The first thing that strikes the observer is an innumerable multitude of men, all equal and alike, incessantly endeavouring to procure the petty and paltry pleasures with which they glut their lives. Each of them, living apart, is as a stranger to the fate of all the rest; his children and his private friends constitute for him the whole of mankind. As for the rest of his fellow-citizens, he is close to them, but he does not see them; he touches them but he does not feel them; he exists only in himself and for himself alone; and if his kindred still remain to him, he may be said at any rate to have lost his country. Above this race of men stands an immense and tutelary power, which takes upon itself alone to secure their gratification and to watch over their fate . . . it is well content that the people should rejoice, provided that they think of nothing but rejoicing.

Democratic societies are tolerant; they do not tell their citizens how they should live, or what will make them happy, virtuous or great. It is not an accident that people in democratic societies are preoccupied with material gain and with the myriad small needs of the body. Nietzsche, who deplored this state of being, said that 'the last man' has 'left the regions where it was hard to live, for one needs warmth'.

> One still works [he goes on] because work is a form of entertainment. But one is careful lest the entertainment be too harrowing. One no longer becomes rich or poor: both require too much exertion. Who still wants to rule? Who to obey? Both require too much exertion. No shepherd and one herd! Everybody wants the same, everybody is the same: whoever feels different goes voluntarily into a madhouse.

Maybe, says Fukuyama, it was boredom which was the underlying cause of the First World War, too much comfort among the bourgeoisie of Europe. If so, they got more discomfort than they bargained for and would not want that again. In 1989 the citizens of the then West Germany were none too excited by the thought of uniting their country when the Wall came down. It might cost too much. It did. Europe's politicians did not hurry to the defence of Bosnia four years later. They knew that there would be no stomach for that cause among their citizens. Only bloodless wars (bloodless, that is, for the democracies) like that of the Gulf, arouse any enthusiasm.

There are no great causes any more. We fill in our résumés in the hope that they may be the pathways to a style of life to which we feel accustomed. It is hard to detect great, unfulfilled longings or irrational passions just beneath the surface of the average first-year law associate. We have to find our pride in sports or eccentricities instead. It may not be great but it is better than any conceivable alternative. We are all last men now.

On the other hand we may not like the end of history when we see it. Fukuyama again: 'Self-interest rightly understood came to be a broadly understandable principle that laid a low but solid ground for public virtue in the United States . . . But in the long run those values had a corrosive effect on the values . . . necessary to sustain strong communities and thereby on a liberal society's ability to be self-sustaining.' Hegel understood that the need to feel pride in one's humanity would not be satisfied by the peace and prosperity that comes with the 'end of history'. In 1806 he wrote, 'We stand at the gates of an important epoch, a time of ferment . . . when a new phase of the spirit is preparing itself.' Almost 200 years later we are in another time of ferment, another dark wood. It may not yet be the end of history.

Maslow was right when he postulated that there was a hierarchy of needs, that when you had enough material goods you moved your sights to social prestige and then to self-realisation. Perhaps, however, his hierarchy did not reach far enough. There could be a stage beyond self-realisation, a stage which we might call idealisation, the pursuit of an ideal or a cause which is more than oneself. It is this extra stage which would redeem the self-centred tone of Maslow's thesis which, for all that it rings true of much of our experience, has a rather bitter aftertaste. Maslow himself was to acknowledge this towards the end of his life.

For the Sake of a Cause

If we are not machines, random accidents in the evolutionary chain, we need to have a sense of direction. Tolstoy, in his confessions, tells how he could find no logical purpose for his existence. He was successful, happily married, rich, yet it all seemed pointless. He came to the conclusion that man only lived because he believed in something. If he didn't believe there was anything there, he would kill himself. Faith, therefore, was 'the power of life'. Laura Ashley, explaining why she started her country fabrics business, said: 'I sensed that most people wanted to raise families, have gardens and live as nicely as they can.' Her business flourished in the Seventies and into the Eighties because I think she caught the mood of the times, the generation of the last men. Mayor Dinkins, however, at Arthur Ashe's memorial service in 1993, said: 'Service to others is the rent we pay for our space on earth. Arthur Ashe paid his rent in full.'

Mayor Dinkins, in his turn, may have caught the mood of the approaching millennium. The RISC survey, mentioned before, identified a growing search for meaning and authenticity as the distinguishing element of the mood of the Nineties, in contrast with the 'boring generation' of the Seventies and Eighties – 'people uninterested in ideological debate and more concerned about themselves'. This new 'ethical' dimension had, they said, several manifestations – 'a sense of purpose, a search for identity, dignity and a quality of life prior to lifestyle (aesthetics and harmony)'.

It is a search for a cause. The cause, however, to be truly satisfying must be a 'purpose beyond oneself', because to be turned in on yourself, said St Augustine, is the greatest of sins; because we discover ourselves through others, said Jung; because the immortality, for which we all privately long, is really immortality through others. This last statement needs some justification because it suggests that most of the religions have got it wrong. There may be some future existence after death, for all that we know, but it will certainly not be expressed in bodily shape, or in time or in space. It is, therefore, literally inconceivable, and, as a result, not something which I myself can take seriously. My purpose in this life, as I read the teachings of the sages, is to so live that others can live better after I have gone, that, if I live on in any sense, I may live on in the continuing lives of others. Heaven and hell I see as medieval forms of social control, along with theories of reincarnation. They are also rather self-centred theories – pie in the sky by and by, if you're good.

We do not need to change the world. To nudge a little bit of it along will be enough. One owner-manager of a bakery once contacted me. 'I want to make my little company the best in the country,' he said. 'What do you mean by best?' I asked. 'Are you talking profits?' 'Only up to a point,' he replied. 'Without some longer-term profitability I won't be able to keep it going, but that's not really the point – I want it to be a showcase, the kind of company of which I and all who work there will be proud to say, "that's my place." ' He had a cause. Art Fury, of Post-It fame, commenting on entrepreneurial success said once, 'Those who invest only to get rich will fail. Those who invest to help others will probably succeed.'

Those who talk about vision as essential for the future of an enterprise are right, but it has to be the sort of vision that others can relate to. Not many in the lower realms of the organisation can get excited by the thought of enriching the shareholders. 'Excellence' and 'quality' are the right sort of words, but they have been tarnished by repetition in too many organisations. They were often synonyms for cost- or people-cutting, or they begged the question – for whom are we doing this? We need to believe in what we are doing if we are to lift ourselves on to a second curve in any enterprise, or if we are going to be prepared to compromise our wishes and our needs for the good of others. Some businesses turn this into a concern for the customer, but we have to wonder whether this concern is not a means rather than an end, a more effective way of doing business.

I once attended a top management seminar arranged by a leading group of hotels. The keynote speech was given by a Benedictine monk who explained St Benedict's view of hospitality. In his monastery, he said, they got many visitors, both men and women, who came for peace and reflection. 'We try to practise St Benedict's command to welcome every man, each man and the whole man.' 'That means,' he explained, ' that we do not discriminate between president and pauper (every man), and we had both last month; we treat each person as an individual (each man), paying attention to their special needs and wishes; lastly, we try to deal with the whole of them, with their deep needs as well as their surface wants, and to enter as fully as they will let us into their lives.' His talk was rapturously received by the executives who saw in it a reason for their hard work, a reason more deeply satisfying than numbers on a balance sheet. When, however, I checked in later to their hotel, I found that every movable item was attached in some way to the walls. Even the toilet-roll in the lavatory was in a locked container. 'We have to,' they explained to me. 'Our visitors will steal anything, given half a chance.' If you can't trust your visitor with the toilet-roll, I reflected, it will be hard to deliver the Benedictine message. Yet for a moment, there, they glimpsed a vision, a direction worth the journey.

As George Bernard Shaw put it, in *Man and Superman*: 'This is the true joy in life, the being used for a purpose recognized by yourself as a mighty one; the being a force of nature instead of a feverish, selfish little clod of ailments and grievances complaining that the world will not devote itself to making you happy.'

Britain will never be 'Great' again, in the sense that she could be a world power or an economic force again, but she could find a new cause and forge a new existence as, for instance, the 'Athens of Europe', meaning the old Athens of learning, culture and the arts. Her great comparative advantage is her language. Everyone, everywhere, wants to learn it. Her universities, theatres, designers, artists, architects, film- and TV-makers, writers and literati, musicians and dancers are world-class. Sadly, she is more likely to be known as a museum than as a cultural centre, but the opportunity is there to find a second curve and to lift her people, to give her a sense of new direction.

Malaysia now boasts a 2020 Vision. The pun is deliberate. It is a 30-year plan outlining the kind of place the country's leaders would like to see it be in 2020. It is underpinned by an optimistic rate of growth – 7.2 per cent, enough to bring it up to American standards of living by 2020 – but that is where it starts, not where it stops. The Vision is full of the ways in which that money will be spent and distributed, on education, on the handicapped, on the old, on the environment (belatedly). Visiting that country I expected cynicism, instead I found excitement. Business leaders had a justification for their efforts. Others had hope. The headlines of the plan were even pinned up in the taxis.

It is hard, in the conditions of comfortable democracy, to find a cause which lifts the efforts of the comfortable ones.

That is why some fear a return to war as a way of putting some energy back into our peoples. Making money not war has turned out to be less inspiring. Another war would be a wasteful way to disprove the end-of-history thesis. It is tempting to call for better leadership, but we probably expect too much from the leaders of the nations. Those nations are too big, the connections not strong enough, the commitment to the future not long enough. It is better to look smaller, to our now-smaller organisations, to local communities and cities, to families and clusters of friends, to small networks of portfolio people with time to give to something bigger than themselves. We have to fashion our own directions in our own places.

A Postscript

Two Stories

My wife's ancestor was Sir Rowland Hill, famed as the inventor of the penny post and the first postage stamps in the 1840s. Until he came on the scene, letters were priced according to their weight and the distance they had travelled, rather logically when you think about it, and were paid for by the recipient. A letter from London to Edinburgh, for instance, might cost one shilling and sixpence, a lot of money in those days, but, then, it was a long distance. Some clever folk used to send an empty envelope to their families, who would then refuse to pay for it on arrival because they had heard what they needed to hear, that their loved one was alive. That put the costs up even more. The result was that only the rich could afford to send real letters to each other. Letter-writing was an élite pastime.

Rowland Hill proposed that we do some upside-down thinking. If every letter cost only one penny, no matter where it went to in Britain, and if it was prepaid by a 'stamp' which you could buy in advance and stick on, he argued that two things would happen: first, the volume of mail would expand hugely, more than compensating for any loss on the cost of the longer deliveries, but, more importantly, everyone would be able to send letters to each other. This would give an enormous boost to education because there would be some practical point in everyone learning to read and write. It would also help the cohesion of the nation because friend would be able to keep in touch with friend, mother with son, wife with distant husband. It would be, he said, not just a commercial success but a significant piece of social reform.

Nobody believed him. It took years of argument and campaigning before he convinced Parliament to make the change. When they did, the results were dramatic. Within 10 years some 50 countries had adopted the idea of pre-bought stamps and the modern postal service was born. Rowland Hill died richly honoured and is remembered to this day as the father of the penny post.

What is interesting about this story, however, is this: when he started his campaigning Rowland Hill was not in the postal service. He was a clerk in the South Australia Commission, having, before that, been a schoolteacher in his father's school. The postal service was nothing to do with him; it was none of his business. He was not rich, nor famous, nor influential, but he cared, he saw something which needed to be done and he decided that he could not live with himself if he didn't do something about it. We can't wait for the mountain to move, we have to climb it ourselves.

We are not, however, all destined to be social reformers. Richard Harries, the Bishop of Oxford, tells another story. There was a rabbi once, called Zuzya of Hannipol. He spent his life lamenting his lack of talent and his failure to be another Moses. One day God comforted him. 'In the coming world,' he said, 'we will not ask you why you were not Moses, but why you were not Zuzya.' We are not gods. We can't do everything, or even very much at all, in the small interval of time we have in this world. It is as much as we can do to be our full selves, full doughnuts and full raincoats.

Fires in the Darkness

There are two photographs on my desk, taken by my wife in South Africa. The first is the head of a small black boy.

He is smiling; everything about his eyes and his face radiates intelligence, enthusiasm, excitement. It is a happy face, full of promise. The second photograph is of the same boy, but this time the photographer has moved back, so that you now see him full-length. You see the shanty hut behind him, his bare feet and the excrement in which he is standing. The two photographs may be a symbol of our challenge today, not only in South Africa. The intelligence and the promise are there, if we can only release them from the chains of their surroundings.

Our people are clever, many of them. Most people are decent, given half a chance. They are not uncaring, if only because they know that a world which crumbles around them will do them no good at all. But first there has to be a general acceptance that the world has changed. The end of communism does not mean that capitalism, in its old form, is therefore the one right way. The triumph of the democracies over totalitarianism does not mean that everything in those democracies is thereby validated. The huge strides made by science in the last decades does not mean that scientists have or could have the answer to everything and that the rest of us need not bother.

It is also the end of the age of the mass organisation, the age when we could all confidently expect to be employed for most our lives if we so wanted, and over 90 per cent did so want. Work will still be central to our lives but we shall now have to rethink what we mean by work and how it might be organised. At first sight, the challenge is daunting, but work in those mass organisations has never been unalloyed bliss for all. The mass organisation has not been with us that long. We should not think of it as a law of nature. Maybe we shall be better off without it.

The hope lies in the unknown, in that second curve, if we can find it. The world is up for reinvention in so many ways. Creativity is born in chaos. What we do, what we belong to, why we do it, when we do it, where we do it –

these may all be different and they could be better. Our societies, however, are built on case law. Change comes from small initiatives which work, initiatives which, imitated, become the fashion. We cannot wait for great visions from great people, for they are in short supply at the end of history. It is up to us to light our own small fires in the darkness.

Bibliography

Abbeglen, James C. and Stalk, George Jun., *Kaisha, the Japanese Corporation*, New York, Basic Books, 1985

Albert, Michel, *Capitalism Against Capitalism*, London, Whurr, 1993

Anderson, Digby (ed.), *The Loss of Virtue*, London, Social Affairs Unit, 1993

Appleyard, Brian, *Understanding the Present*, London, Pan Books, 1991

Baden-Fuller, Charles and Stopford, John, *Rejuvenating the Mature Corporation*, London, Routledge, 1992

Bahrami, Homa, 'The Emerging Flexible Organization', *California Management Review*, Summer 1992

Ball, Christopher, 'The Adelphi Idler', London, *RSA Journal*, May 1993

Bennis, Warren, *An Invented Life*, Reading, Mass., Addison Wesley, 1993

Bishop, Jeff and Hoggett, Paul, *Organizing Around Enthusiasms*, London, Comedia, 1988

Bloom, Allan, *The Closing of the American Mind*, New York, Simon and Schuster, 1987

Commission for Social Justice, *The Justice Gap*, London, IPPR, 1993

Drucker, Peter, *Post-Capitalist Society*, Oxford, Butterworth-Heinemann, 1993

Fukuyama, Francis, *The End of History and the Last Man*, London, Hamish Hamilton, 1992

Galbraith, John K., *The Culture of Contentment*, London, Sinclair Stevenson, 1992

Gorz, P., *A Critique of Economic Reason*, London, Verso, 1989

Goyder, George, *The Just Enterprise*, London, André Deutsch, 1987
Hammer, Michael and Champy, James, *Re-engineering the Corporation*, New York, Harper Collins, 1993
Hampden-Turner, Charles, *Corporate Culture*, London, Hutchinson, 1990
Hampden-Turner, Charles, *The Seven Cultures of Capitalism*, London, Piatkus Books, 1994
Havel, Vaclav, *Disturbing the Peace*, New York, Vintage Books, 1991
Hegel, G., *The Philosophy of History*, London, Dover Publications, 1956
Henzler, H.A., 'Eurocapitalism', Harvard Business Review, Jul/Aug. 1992
Hewitt, Patricia, *About Time*, London, Rivers Oram Press, 1993
Kanter, Rosabeth M., *When Giants Learn to Dance*, London, Simon and Schuster, 1989
Keegan, William, *The Spectre of Capitalism*, London, Radius, 1993
Kennedy, Paul, *Preparing for the Twenty-First Century*, New York, Random House, 1993
Kester, W. Carl, *Japanese Takeovers*, Boston, Harvard Business School Press, 1991
Kraus, Michael, *The End of Equality*, New York, Basic Books, 1992
Leinberger, Paul and Tucker, Bruce, *The New Individualists*, New York, Harper Collins, 1991
Lucas, J.R., *On Justice*, Oxford, Clarendon Press, 1980
Nietzsche, F., *Beyond Good And Evil*, New York, Vintage Books, 1968
O'Neil, John R., *The Paradox of Success*, New York, Putnam, 1993
Osborne, David and Gaebler, Ted, *Re-inventing Government*, Reading, Mass., Addison Wesley, 1992
Peters, Tom, *Liberation Management*, New York, Knopf, 1992

Rauch, Jonathan, *The Outnation*, Boston, Harvard Business School Press, 1992

Reich, Robert, *The Work of Nations*, New York, Knopf, 1991

Sampson, Anthony, *The Essential Anatomy of Britain*, London, Hodder and Stoughton, 1993

Schor, Juliet B., *The Overworked American*, New York, Basic Books, 1992

Schumacher, E.F., *Small is Beautiful*, London, Blond & Briggs Ltd, 1973

Schwartz, Peter, *The Art of the Long View*, New York, Doubleday, 1991

Semler, Ricardo, *Maverick*, London, Hutchinson, 1993

Senge, Peter, *The Fifth Discipline*, New York, Doubleday, 1990

Shames, Laurence, *The Hunger for More*, New York, Times Books, 1989

Stayer, Ralph, 'How I Learnt to Let My Workers Lead', Harvard Business Review, Nov/Dec. 1990

Stewart, Rosemary, *Choices for the Manager*, London, McGraw Hill, 1983

Thurow, Lester, *Head to Head*, New York, William Morrow, 1992

Trompenaars, Alfons, 'The Organization of Meaning and the Meaning of Organizations', doctoral dissertation, Wharton School, 1987

Waldrop, M. Mitchell, *Complexity*, New York, Simon and Schuster, 1992

Watkinson Report, *The Responsibility of the British Public Company*, London, British Institute of Management, 1972

Young, Michael, *The Rise of the Meritocracy*, London, Penguin, 1961

Index